S0-CAS-462

Seventy Years in Organic Chemistry

Tetsuo Nozoe

PROFILES, PATHWAYS, AND DREAMS
Autobiographies of Eminent Chemists

Jeffrey I. Seeman, Series Editor

American Chemical Society, Washington, DC 1991

CHEM.

Library of Congress Cataloging-in-Publication Data

Nozoe, Tetsuo, 1902–
Seventy years in organic chemistry/Tetsuo Nozoe
p. cm.–(Profiles, pathways, and dreams: autobiographies of eminent chemists, ISSN 1047–8329)
Includes bibliographical references (p.) and index.
ISBN 0–8412–1769–6 (cloth).–ISBN 0–8412–1795–5 (paper)
1. Nozoe, Tetsuo, 1902– . 2. Chemists—Japan—Biography. 3. Chemistry, Organic—Japan—History—20th century.

I. Title. II. Title:Seventy years in organic chemistry. III. Series: Profiles, pathways, and dreams

QD22.N75A3 1991 540'.092—dc20 [B] 90–876

Jeffrey I. Seeman, Series Editor

The paper used in this publication meets the minimum requirements of American National Standard for Information Sciences—Permanence of Paper for Printed Library Materials, ANSI Z39.48–1984.

∞

PRINTED IN THE UNITED STATES OF AMERICA

1991 ACS Books Advisory Board

Foreword

In 1986, the ACS Books Department accepted for publication a collection of autobiographies of organic chemists, to be published in a single volume. However, the authors were much more prolific than the project's editor, Jeffrey I. Seeman, had anticipated, and under his guidance and encouragement, the project took on a life of its own. The original volume evolved into 22 volumes, and the first volume of Profiles, Pathways, and Dreams: Autobiographies of Eminent Chemists was published in 1990. Unlike the original volume, the series was structured to include chemical scientists in all specialties, not just organic chemistry. Our hope is that those who know the authors will be confirmed in their admiration for them, and that those who do not know them will find these eminent scientists a source of inspiration and encouragement, not only in any scientific endeavors, but also in life.

M. Joan Comstock
Head, Books Department
American Chemical Society

Contributors

We thank the following corporations and Herchel Smith for their generous financial support of the series Profiles, Pathways, and Dreams.

Akzo nv

Bachem Inc.

E. I. du Pont de Nemours
and Company

Duphar B.V.

Eisai Co., Ltd.

Fujisawa Pharmaceutical Co., Ltd.

Hoechst Celanese Corporation

Imperial Chemical Industries PLC

Kao Corporation

Mitsui Petrochemical Industries,
Ltd.

The NutraSweet Company

Organon International B.V.

Pergamon Press PLC

Pfizer Inc.

Philip Morris

Quest International

Sandoz Pharmaceuticals
Corporation

Sankyo Company, Ltd.

Schering–Plough Corporation

Shionogi Research Laboratories,
Shionogi & Co., Ltd.

Herchel Smith

Suntory Institute for Bioorganic
Research

Takasago International
Corporation

Takeda Chemical Industries, Ltd.

Unilever Research U.S., Inc.

Profiles, Pathways, and Dreams

Titles in This Series

Sir Derek H. R. Barton *Some Recollections of Gap Jumping*

Arthur J. Birch *To See the Obvious*

Melvin Calvin *Following the Trail of Light: A Scientific Odyssey*

Donald J. Cram *From Design to Discovery*

Michael J. S. Dewar *A Semiempirical Life*

Carl Djerassi *Steroids Made It Possible*

Ernest L. Eliel *From Cologne to Chapel Hill*

Egbert Havinga *Enjoying Organic Chemistry, 1927–1987*

Rolf Huisgen *Mechanisms, Novel Reactions, Synthetic Principles*

William S. Johnson *A Fifty-Year Love Affair with Organic Chemistry*

Raymond U. Lemieux *Explorations with Sugars: How Sweet It Was*

Herman Mark *From Small Organic Molecules to Large: A Century of Progress*

Bruce Merrifield *The Concept and Development of Solid-Phase Peptide Synthesis*

Teruaki Mukaiyama *To Catch the Interesting While Running*

Koji Nakanishi *A Wandering Natural Products Chemist*

Tetsuo Nozoe *Seventy Years in Organic Chemistry*

Vladimir Prelog *My 132 Semesters of Chemistry Studies*

John D. Roberts *The Right Place at the Right Time*

Paul von Rague Schleyer *From the Ivy League into the Honey Pot*

F. G. A. Stone *Organometallic Chemistry*

Andrew Streitwieser, Jr. *A Lifetime of Synergy with Theory and Experiment*

Cheves Walling *Fifty Years of Free Radicals*

About the Editor

JEFFREY I. SEEMAN received his B.S. with high honors in 1967 from the Stevens Institute of Technology in Hoboken, New Jersey, and his Ph.D. in organic chemistry in 1971 from the University of California, Berkeley. Following a two-year staff fellowship at the Laboratory of Chemical Physics of the National Institutes of Health in Bethesda, Maryland, he joined the Philip Morris Research Center in Richmond, Virginia, where he is currently a section leader. In 1983–1984, he enjoyed a sabbatical year at the Dyson Perrins Laboratory in Oxford, England, and claims to have visited more than 90% of the castles in England, Wales, and Scotland.

Seeman's 80 published papers include research in the areas of photochemistry, nicotine and tobacco alkaloid chemistry and synthesis, conformational analysis, pyrolysis chemistry, organotransition metal chemistry, the use of cyclodextrins for chiral recognition, and structure–activity relationships in olfaction. He was a plenary lecturer at the Eighth IUPAC Conference on Physical Organic Chemistry held in Tokyo in 1986 and has been an invited lecturer at numerous scientific meetings and universities. Currently, Seeman serves on the Petroleum Research Fund Advisory Board. He continues to count Nero Wolfe and Archie Goodwin among his best friends.

Contents

Photographs

Preface

"HOW DID YOU GET THE IDEA—and the good fortune—to convince 22 world-famous chemists to write their autobiographies?" This question has been asked of me, in these or similar words, frequently over the past several years. I hope to explain in this preface how the project came about, how the contributors were chosen, what the editorial ground rules were, what was the editorial context in which these scientists wrote their stories, and the answers to related issues. Furthermore, several authors specifically requested that the project's boundary conditions be known.

As I was preparing an article[1] for *Chemical Reviews* on the Curtin–Hammett principle, I became interested in the people who did the work and the human side of the scientific developments. I am a chemist, and I also have a deep appreciation of history, especially in the sense of individual accomplishments. Readers' responses to the historical section of that review encouraged me to take an active interest in the history of chemistry. The concept for Profiles, Pathways, and Dreams resulted from that interest.

My goal for Profiles was to document the development of modern organic chemistry by having individual chemists discuss their roles in this development. Authors were not chosen to represent my choice of the world's "best" organic chemists, as one might choose the "baseball all-star team of the century". Such an attempt would be foolish: Even the selection committees for the Nobel prizes do not make their decisions on such a premise.

The selection criteria were numerous. Each individual had to have made seminal contributions to organic chemistry over a multidecade career. (The average age of the authors is over 70!) Profiles would represent scientists born and professionally productive in different countries. (Chemistry in 13 countries is detailed.) Taken together, these individuals were to have conducted research in nearly all sub-specialties of organic chemistry. Invitations to contribute were based on solicited advice and on recommendations of chemists from five continents, including nearly all of the contributors. The final assemblage was selected entirely and exclusively by me. Not all who were invited chose to participate, and not all who should have been invited could be asked.

A very detailed four-page document was sent to the contributors, in which they were informed that the objectives of the series were

1. to delineate the overall scientific development of organic chemistry during the past 30–40 years, a period during which this field has dramatically changed and matured;

2. to describe the development of specific areas of organic chemistry; to highlight the crucial discoveries and to examine the impact they have had on the continuing development in the field;

3. to focus attention on the research of some of the seminal contributors to organic chemistry; to indicate how their research programs progressed over a 20–40-year period; and

4. to provide a documented source for individuals interested in the hows and whys of the development of modern organic chemistry.

One noted scientist explained his refusal to contribute a volume by saying, in part, that "it is extraordinarily difficult to write in good taste about oneself. Only if one can manage a humorous and light touch does it come off well. Naturally, I would like to place my work in what I consider its true scientific perspective, but . . ."

Each autobiography reflects the author's science, his lifestyle, and the style of his research. Naturally, the volumes are not uniform, although each author attempted to follow the guidelines. "To write in good taste" was not an objective of the series. On the contrary, the authors were specifically requested not to write a review article of their field, but to detail their own research accomplishments. To the extent that this instruction was followed and the result is not "in good taste", then these are criticisms that I, as editor, must bear, not the writer.

As in any project, I have a few regrets. It is truly sad that Egbert Havinga, who wrote one volume, and David Ginsburg, who translated another, died during the development of this project. There have been many rewards, some of which are documented in my personal account of this project, entitled "Extracting the Essence: Adventures of an Editor" published in *CHEMTECH*.[2]

Acknowledgments

I join the entire chemical community in offering each author unbounded thanks. I thank their families and their secretaries for their contributions. Furthermore, I thank numerous chemists for reading and reviewing the autobiographies, for lending photographs, for sharing information, and for providing each of the authors and me the encouragement to proceed in a project that was far more costly in time and energy than any of us had anticipated.

I thank my employer, Philip Morris USA, and J. Charles, R. N. Ferguson, K. Houghton, and W. F. Kuhn, for without their support Profiles, Pathways, and Dreams could not have been. I thank ACS Books, and in particular, Robin Giroux (acquisitions editor), Karen Schools Colson (production manager), Janet Dodd (senior editor), Joan Comstock (department head), and their staff for their hard work, dedication, and support. Each reader no doubt joins me in thanking 24 corporations and Herchel Smith for financial support for the project.

I thank my children, Jonathan and Brooke, for their patience and understanding; remarkably, I have been working on Profiles for more than half of their lives—probably the only half that they can remember! Finally, I again thank all those mentioned and especially my family, friends, colleagues, and the 22 authors for allowing me to share this experience with them.

JEFFREY I. SEEMAN
Philip Morris Research Center
Richmond, VA 23234

November 11, 1990

[1] Seeman, J. I. *Chem. Rev.* **1983,** *83,* 83–134.

[2] Seeman, J. I. *CHEMTECH* **1990,** *20*(2), 86–90.

Editor's Note

TETSUO NOZOE IS A MAN OF BOUNDLESS energy and great mental alertness—amazingly so for a man who was honored last year for attaining his *Beiju.** Recently, I asked Nozoe two questions: "What drives a man in his 80s to accomplish, to produce at such a demanding pace? Why do you write about your latest results in such detail?"

Nozoe replied that he was afraid he might not live long enough to complete the publication of the research in this very curious area (troponoid chemistry). If I describe it, "anyone who might have an interest could be expected to develop it further." Because of his involvement in writing this book, "to my great pleasure, my co-workers offered a great deal of cooperation, and some 20 papers were published during these two years."

Nozoe wrote the entire book, including the nonchemical part of the text, in longhand and in Japanese. Structures, reactions, and novel products flowed from his pen, as did personal observations, characterizations of his colleagues and students, scientific philosophies, and his views of Japanese organic chemistry. But all this did not come as easily as it sounds. Japanese culture stresses the accomplishments of the group versus recognition of the individual—a philosophy diametrically opposed to the requirements for an autobiography. However, according to Nozoe, "the situation is gradually changing as the Japanese scientific community becomes internationalized."

**A* beiju *is a celebration of attaining the 88th birthday. See page 105 for a more detailed explanation.*

As were all contributors to the *Profiles* series, Nozoe was asked to place his research accomplishments within the context of his personal life, to ruminate on his philosophic beliefs, and to reminisce about his friends and life experiences. This he did, graciously and in rich detail, although it was difficult for him to overcome his natural inclination to self effacement. Reflecting on the changing nature of tradition, in this instance dialogue with fellow chemists, Nozoe observed, "In my era, we were much more timid and neither asked questions nor voiced our opinions, especially with foreign speakers. Probably this is due to a special old Japanese characteristic . . ."

Scientists seldom document the personal side of their lives as Nozoe has done. This volume is enhanced by signatures, inscriptions, and drawings from his autograph books (*see* p 113). In his seventy years of work and travel, Nozoe has recorded his meeting of many of the chemists who have become friends and whose names are so well known in this field. He recalled Vladimir Prelog saying to him that his "autograph books are very precious, and I must keep them in a museum, but if I ever be[come] very poor, I would become [a] millionaire if I sell the autograph book cut by scissors piece by piece." It is doubtful whether another such collection even exists!

Nozoe's friends abound. His co-worker Professor Hiroshi Yamamoto edited and translated this book with great devotion, and Koji Nakanishi provided the English translation of Nozoe's handwritten conclusion at the risk of missing the deadline for his own *Profiles* book. Concerned for his friend, Nozoe told me, "I told Professor Nakanishi to send in his [own] draft soon, because he always looks extremely busy." But Nakanishi has always found time and energy for his good friend Nozoe. Indeed, forty years ago, when Nozoe first traveled to Europe and the United States, it was Nakanishi who taped the English translation of Nozoe's lecture so that Nozoe could practice his English for his presentations.

Very few organic chemists outside Japan have extensive knowledge or understanding of the splendid historical and educational relationships in Japanese chemistry. Japanese contributions to organic chemistry have steadily increased in significance and prominence internationally, hence the need for a greater understanding of this heritage. With that in mind, Nozoe devised the "Majima organic chemistry tree", which illustrates Japanese academic connections starting with the early 20th century chemist–scholar, Riko Majima. According to Nakanishi, "the homogeneous nature of Japanese society made the selection of names by Nozoe a rather delicate issue." Nozoe, a thoughtful and sensitive man, did not wish to offend his colleagues and so was reluctant to publish it. I urged him to do so because of its historical value and for the reader's benefit. He concurred.

Without any question, Nozoe has had a major impact on organic chemistry in general and on his country's science in particular. Together with Dewar's structure for stipitatic acid and Erdtman's structures for the thujaplicins, Nozoe's tropolone structure determination and subsequent chemistry were responsible for the beginning of a new field of chemistry: novel aromatic systems. In addition, Nozoe's personality and chemical successes for seven decades—70 years!—have set the example and forged the leadership that helped bring Japanese organic chemistry to the heights it now enjoys.

Nozoe's current goal is to write his "long-planned Comprehensive Tropylium Chemistry (~1500 pages). This is the most important task I still have to do. In order to accomplish this task within the next two years, I have rented a room in a hotel, which is situated about 15 minutes by taxi from both the Kao Tokyo Research Laboratory and the Tokyo Station. Until now it took me about 2½ to 3 hours, sometimes 4 hours, to commute between my house and office in Kao by taxi every day. Time is too precious to an old man like me to waste, so I decided to rent the room. I must by all means live to see this project completed." Recently, Nozoe returned to his home to complete his task, as he found working in a small hotel room very inconvenient.

Nozoe is a model of integrity, hard and dedicated work, inspiration, loyalty, and devotion. He is also an editor's dream. Nozoe wrote to me: "The proofreading is expected to take [a] rather long time for me [as an] elder person, so that I would like to get the draft as soon as possible." He does not wait for tasks; he seeks them out. I wish my children could experience Tetsuo Nozoe as I have. He is a model for us all.

Jeffrey I. Seeman
September 12, 1991
Richmond, Virginia

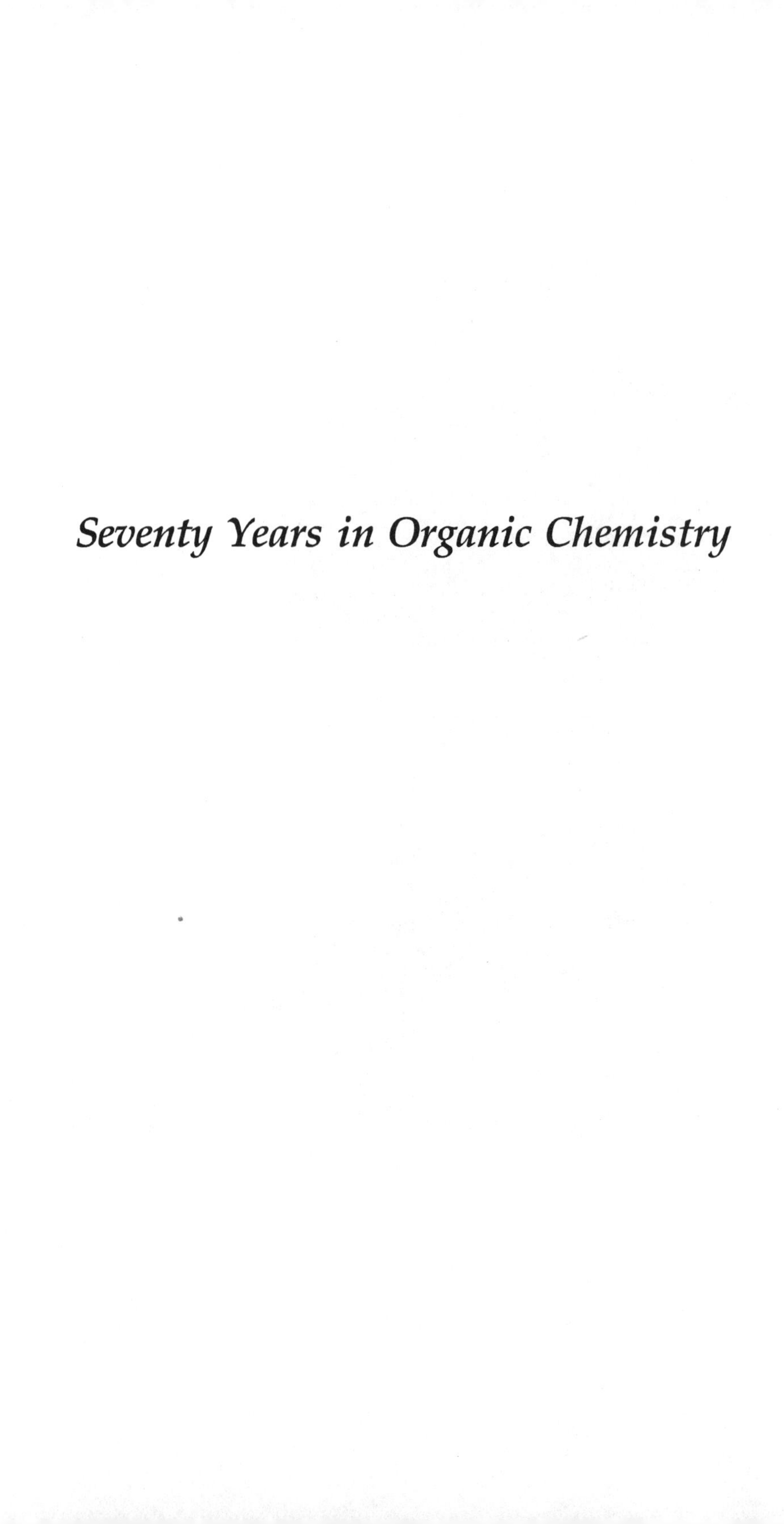

Seventy Years in Organic Chemistry

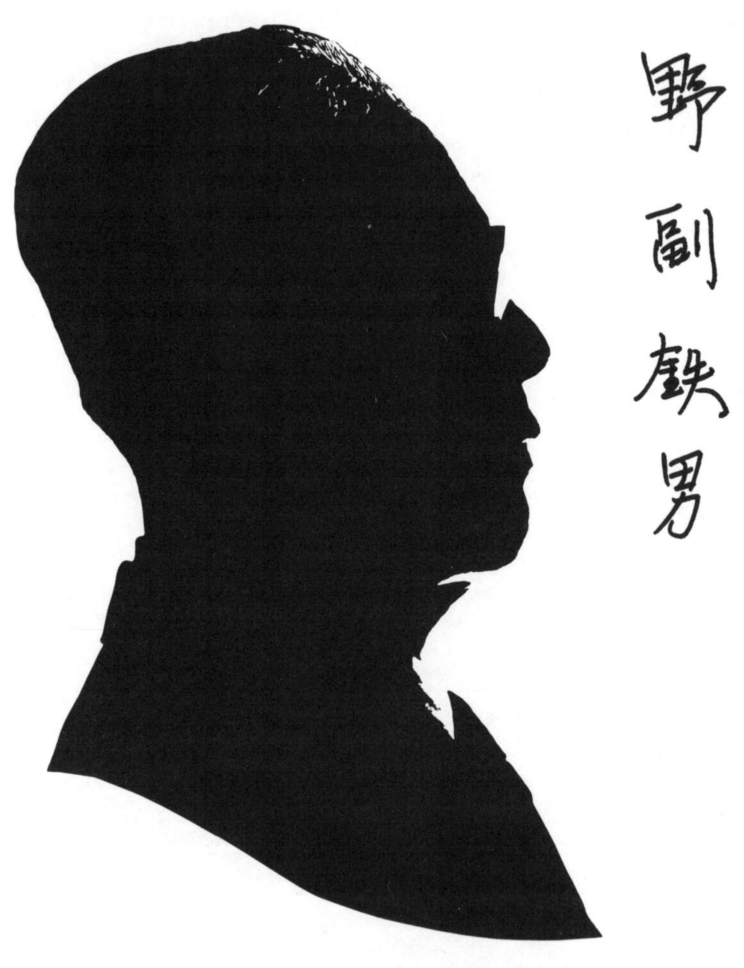

Tetsuo Nozoe

My Beginnings

Early Days (1902–1923)

I was born the sixth son of Juichi and Toyoko Nozoe on May 16, 1902, in Sendai City, Japan. I had seven brothers and three sisters, and among the siblings in the family, I was the middle child. At that time, large families were not uncommon; however, our family of 11 children was still considered larger than normal.

Both of my parents were born in Nagasaki Prefecture, the southernmost part of Japan. After graduating from Tokyo Imperial University Law School, my father settled in Sendai City to practice law. My father was very busy as an attorney and a politician; hence, he could not afford to spend much time with us. He was involved in politics and at one time served as a chairman of a municipal assembly, as well as a member of the House of Representatives. My mother was a devout Christian. She enrolled all of her daughters in an American missionary school in Sendai and had them study piano in the special music course offered there.

My parents wanted each of their sons to pursue a different occupation. My eldest brother practiced law in Tokyo, but unfortunately, he contracted a fatal disease and died at a relatively young age. Two other brothers also passed away before they finished their university studies. Of my four remaining brothers, two became engineers for overseas trading companies, the third became a professor and, later, a principal of an agricultural college in Manchuria, and the fourth established himself as a banker, also in Manchuria. Of the 11 children in my family, only my youngest sister and I are still living, even though I was in very poor health throughout my childhood.

As a schoolboy, my friends and I often played a "spelunking game". In this game, we investigated the caves in a nearby hill with a flashlight and mapped out our "journey". After I entered the fourth grade (primary school), I was stricken on successive occasions with bouts of dysentery, typhoid, beriberi, and pneumonia. Because recovery from such diseases took a long time in those days, I missed school often during the 4 years from the fourth to the eighth grade (middle school). For fear that I would have to repeat the same classes, my mother employed a private tutor for me. These 4 years during which I was often absent from school might be the reason why I was not, and still am not, a competent writer!

After I entered junior high school, I became interested in philately. I joined the International Philatelic Society and began corresponding with pen pals abroad in Esperanto and exchanged

Tetsuo Nozoe on his second birthday, May 16, 1904, in Sendai. He holds the Japanese naval ensign, a reflection of the influence of militarism in those days.

stamps with them. Over the course of time, I gradually regained my health, and my grades improved.

Although my parents wanted me to be a doctor, I found chemistry to be interesting and thus decided that I wanted to become a chemist. I asked my parents to allow me to use a section of the storage shed in the corner of the garden as my laboratory. After obtaining their permission, I set up my experiment bench; bought glassware, chemicals, and an alcohol lamp from a nearby pharmacy; and attempted experiments of inorganic and organic chemistry that could only be likened to childish mischief. My mother was very worried that I might cause a fire or be hurt. If the "garden laboratory" of my middle-school period is included, my relation with chemistry extends to over 70 years. Because I enjoyed biology but disliked engineering graphics, I entered the premedical class at what was known in the old Japanese education system as the second higher school in Sendai. (The first higher school was in Tokyo, and the third was in Kyoto.) The chemistry lectures at our second higher school were often too descriptive and did not offer any experiments during the course of the lectures. Thus these chemistry courses did not completely attract my interest.

At that time, the world-famous Professor Hans Molisch from Austria was teaching in the department of biology at Tohoku Imperial

A family portrait taken at home in Sendai, 1910. Tetsuo Nozoe was one of 11 children. First row from left: Shigekatsu (the 7th child in the family); Yaeko (the 8th); his father, Juichi; Fumiko (the 10th); his mother, Toyoko, holding Shizuko (the 11th); Masao (the 9th); and Tetsuo (the 6th). Second row from left: Yasukuni (the 5th); Shigemasa (the 3rd); Satoshi (the 1st); and Kenji (the 2nd). The 4th child, Tsunenobu, was deceased.

University as a visiting scientist. Because he was an acquaintance of my father, I was, on one occasion, able to visit Professor Molisch in his laboratory, where he showed me some of his experiments, for example, the color reaction of sugars (the Molisch reaction). Also, when the distinguished Nobel Prize winner in physics, Dr. Albert Einstein, visited Sendai in 1922 to give a public lecture, I pleaded with my father to arrange for me to attend his lecture. Through Professor Keiichi Aichi of Tohoku Imperial University, who was a friend of both Dr. Einstein and my father, I met Dr. Einstein and was even given his autograph. Although the contents of the lecture were beyond my comprehension, I was, nevertheless, ecstatic over my introduction to Dr. Einstein. At this stage, I had no desire to become a doctor and continued to be interested in organic chemistry. Finally, my parents agreed to let me study chemistry in the university.

At this time I also developed a passion for Western classical music. In Japan, records of classical music were not yet manufactured; the only ones available were the imported single-faced records, "His Master's Voice". I bought one record after another, from symphony music, piano and violin concertos, and violin solos to Italian opera and from Beethoven and Mozart to Caruso and Gallicurci. Sometimes I even listened to one of my friend's records on the phone at midnight! When a famous musician from Europe or the United States visited Japan, I would travel to Tokyo to see his performance, undaunted by the fact that the journey was 12 hours each way.

During this era of my life, my mother spoke to one of her friends from church, Professor Masao Katayama, who was a professor of physical chemistry at Tohoku Imperial University. On hearing that I wanted to study organic chemistry, he immediately recommended Tohoku Imperial University because they had on their staff Professor Riko Majima, the foremost organic chemist in our country at that time. In those days, the national universities of Japan were called imperial universities to distinguish them from the private universities. Tohoku Imperial University is the third oldest imperial university, after Tokyo University (the oldest) and Kyoto University (the second oldest). Some noted private universities also existed at the time; however, their science- and engineering-related departments were poorly equipped because the facilities for such departments were very expensive. Tokyo University was generally considered the best; however, Tohoku Imperial University opened its doors not only to graduates of the higher school (under the old educational system) but also to those who had graduated from industrial, pharmaceutical, or agricultural colleges; to students of higher normal schools; and to females.

Tohoku Imperial University (1923–1926)

In April 1923, I entered the Department of Chemistry of the Faculty of Science at Tohoku Imperial University. At the time, Professor Katayama accepted a position in the chemistry department at Tokyo University. Since that year, the physical chemistry course at Tokyo University flourished under Professor Katayama, although his absence from Tohoku Imperial University was definitely noticed. On a more positive note, however, the organic chemistry department at Tohoku Imperial University ranked as the finest in Japan.

My university was just a 10-minute walk from our home. Since the university's establishment in 1907, an important priority had been placed on creative research. Laboratories and libraries remained open at midnight, not only to researchers but also to students who continued studying after class. After I entered the university, I was able to indulge myself in experimental procedures in chemistry at the university far into the night. I spent less and less time with my hobbies of music and philately. Instead, organic chemistry took up all of my time and became my passion. Almost all of the students at Tohoku Imperial University were Japanese, although there were a few Chinese and Korean students as well. In the chemistry department, 15 undergraduate students were registered in my class. Because some students entered the university after having been employed for some time, the ages of the students varied, with the oldest being about 40.

Many of the science professors in Japan, especially those in organic chemistry, had studied in Europe for a number of years. Thus, much of the research apparatus and methods were often of the European style. The first-year curriculum of general organic chemistry given by Professor Majima was especially enlightening. The chemical compounds (mainly the reagents of Kahlbaum from Germany) to which the professor referred in his lectures were often circulated during the class. Professor Majima's assistant, who was exclusively in charge of the experiments during the organic chemistry lectures, prepared all the major experiments for a lecture the previous day so that the reactions were already in progress in the lecture hall the next morning. We observed such advanced experiments as the Grignard reaction, the formation of the free triphenylmethyl radical and its autoxidation in air, and the formation and spontaneous combustion of dimethyl zinc. The texts and reference books we used for organic chemistry were all written in German.

In those days at the imperial universities, a *kôza* (derived from the German word *Lehrstuhl*) or research system was adopted. A group of staff members, including the head professor, an assistant professor, sometimes a lecturer, and two assistants, was in charge of the laboratories and the undergraduate research. A postgraduate course existed; however, only a few students were enrolled in this program. At that time, it was possible to apply for a doctorate degree based on research performed while assisting one's professor, or even while performing research within a company. The restriction to this practice was that the application for the degree could be submitted only after a level of study higher than that obtained by students in the graduate program had been completed. Thus, a doctorate in science often took more than 10 years to complete.

In the department of chemistry at Tohoku Imperial University, as well as at other imperial universities, at least four research areas (*kôza*) were offered: analytical chemistry, theoretical chemistry, inorganic chemistry, and organic chemistry. Biochemistry was sometimes also included. Professor Majima's organic chemistry *kôza* was the largest in all of Japan. His assistant professor, Dr. Hiroshi Nomura, headed the biochemistry section, which was an independent group of seven or eight people. Professor Majima's group consisted of a lecturer, two assistants, and a subassistant, as well as those individuals at the Sendai branch of his laboratory in the Institute of Physical and Chemical Research (Riken) in Tokyo. With the paid and unpaid researchers, along with one or two postgraduate students, the group consisted of around 20 members.

In those days, students in their final year were required to spend 1 year conducting research for their undergraduate thesis under the supervision of a professor. I selected Professor Majima as my supervisor. Including myself, four senior students were preparing their undergraduate theses under his direction. I was granted a place in a small laboratory under the direct control of Professor Majima, along with Harusada Suginome (a lecturer) and Dr. Shinichi Morio (an assistant), who were studying aconite alkaloids. The other three students each

Graduation day from Tohoku Imperial University in Sendai, 1926. This photo includes classmates and staff. First row from left: I. So, Y. Matsuike (lecturer, physical chemistry), H. Suginome (lecturer, organic chemistry), H. Nomura (assistant professor, biochemistry), M. Ogawa (president, Tohoku Imperial University, inorganic chemistry), Riko Majima (professor, organic chemistry), M. Kobayashi (professor, analytical chemistry), F. Ishikawa (professor, inorganic chemistry), S. Mizukuri (professor, theoretical chemistry), T. Ogata (lecturer, organic chemistry), and Y. Kawakami. Second row from left: S. Ogawa, S. Chikamori, Tetsuo Nozoe, T. Kato, S. Furukawa, E. Shibata, K. Iwamoto, M. Oku, T. Fukae, J. Iwao, and M. Murakami. Circles from left: I. Yoshida and S. Nagami (lecturer, inorganic chemistry). (Photo courtesy of I. Murata.)

carried out their studies under the guidance of their respective supervisors (who were Professor Majima's assistants). The main research topics pursued by Professor Majima's group at that time were structural studies of various aconite alkaloids and other natural products and the syntheses and reactions of a variety of indole derivatives and amino acids. In addition to these topics, industrial projects, such as indole synthesis from acetylene and aniline and synthetic fuel from fish oil, were undertaken.

Unexpectedly, Professor Majima proposed the synthesis of the thyroid hormone, thyroxine, for my thesis project because at the time, structure **1** had been (incorrectly) assigned[1] to thyroxine. It seemed almost impossible for me to synthesize such a compound as **1**. However, using the readily available quinoline-2,4-dicarboxylic acid, **2**, I intended first to prepare compound **3** (parent structure of **1**) by reductive ring opening and recyclization. The first year passed very quickly

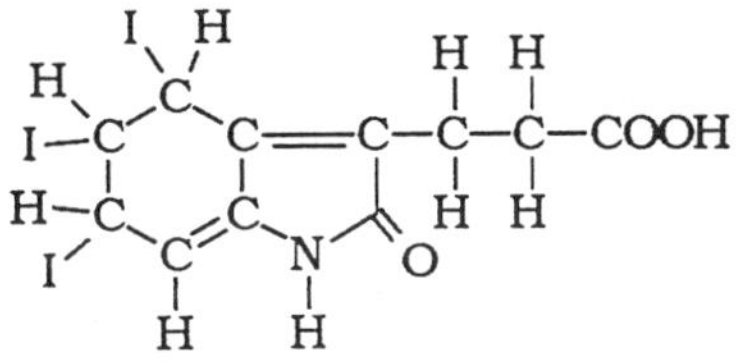

1

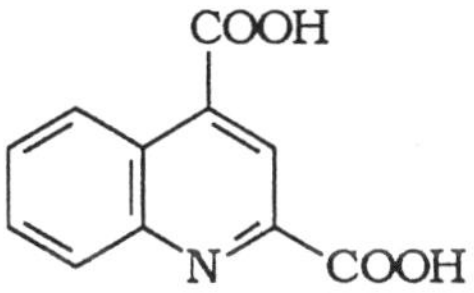

2

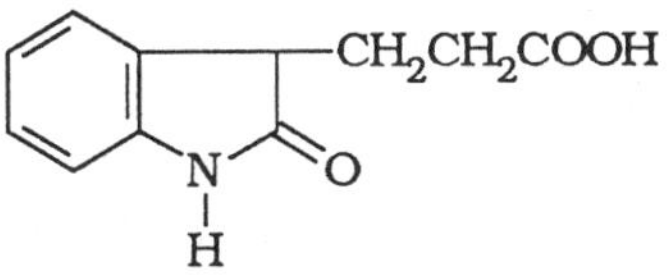

3

as I tried various synthetic approaches. Although Professor Majima thought that the formula **1** for thyroxine might be incorrect, he seemed to think it worthwhile to have me attempt the synthesis of this compound for a year on my own initiative.

Indeed, during my final year at Tohoku Imperial University, I was rarely able to receive Professor Majima's direct instruction regarding my research problems because of his frequent official trips and busy schedule, including several months of overseas travel. Even so, as an enthusiastic Christian, he taught us that one should always conduct research with an infant's curiosity, open-mindedness, and a humble attitude. Such advice, as well as the diligence of the laboratory members who worked day and night, had a profound influence on the rest of my professional career. Among those with whom I worked in the Majima laboratory, nine went on to become professors at other universities.

Two of these people, Harusada Suginome (Hokkaido University) and Shiro Akabori (Osaka University), along with Professor Riko Majima (Osaka University) himself, went on to serve as president of one of the imperial universities.

After graduation from Tohoku Imperial University and at the suggestion of Professor Majima, I compiled my undergraduate research results and submitted them, in Japanese, to *Nippon Kagaku Kaishi (Journal of the Chemical Society of Japan)* under my own name; it was published in 1927. At the same time, Professor Majima summarized my thesis in German and kindly submitted it to the *Proceedings of the Imperial Academy*. This paper, entitled *Über die Reduktion von Chinolindicarbonsäuren*, was published in 1926[2] and was to be the first of many publications. On seeing this first paper of mine, I was extremely happy to be considered a chemist.

Usually, a student's thesis was published jointly with his supervisory professor; thus, to publish a student's first piece of undergraduate research under his own name is exceptional. One possible reason why Professor Majima granted me this privilege was that the thyroxine work had not been in his area of research, and as well, he was perhaps encouraging my research abilities. Immediately after graduation, I was offered the position of paid subassistant at Professor Majima's laboratory. I was preparing to resume my research when Professor Majima approached me and strongly urged me to go to Formosa (now Taiwan) instead to fill a vacant post at the Government Monopoly Bureau Research Laboratories in Taipei. Taiwan (or, as it was known at that time, Formosa) had just decided to establish an imperial university. After some consideration, I finally accepted his recommendation, which, as it turned out unexpectedly, resulted in my 22-year stay in Formosa.

Journey to Formosa

When I left Sendai for Taipei on May 28, 1926, Professor Majima, who came to the railway station to see me off, informed me that the correct structure for the deiodo derivative, **4**, of thyroxine, $C_{15}H_{11}O_4NI_4$, had just been published by Harrington.[3] I was surprised to see the new formula **4**, because it was quite different from the skeleton of formula **1**,

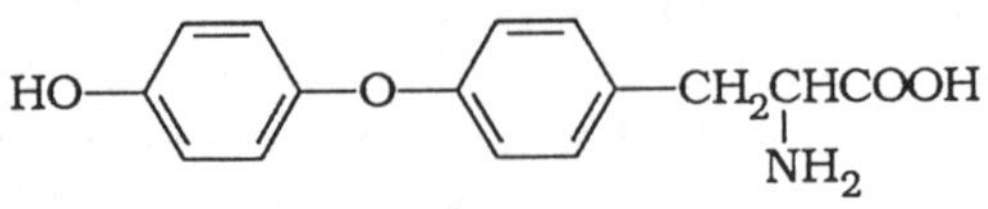

HO–⟨C₆H₄⟩–O–⟨C₆H₄⟩–$CH_2CH(NH_2)COOH$

4

with which I was familiar. Kendal[1] had succeeded in isolating only 7 mg of the hormone, a very small amount for structural elucidation in those days, and the compound contained too much iodine to allow determination of its correct formula. The formula 1 proposed by Kendal was based on his speculation, which was inevitable in those early days.

My journey to Taipei took 5 days by train and ship from Sendai. I was not at all hesitant to move to Formosa, even though I had rarely been outside of my prefecture since my birth. My parents were from Nagasaki Prefecture, and all my relatives lived there, so I did not hold so much attachment to my birthplace. Perhaps the thought of discovering new and interesting natural products in tropical lands encouraged my decision to leave. When my parents learned that I was going to Formosa, they were very surprised; my mother, especially, was strongly opposed to it. However, after talking with Professor Majima, both my parents accepted my decision. Even so, neither they nor I dreamed that I would live in Formosa for more than 20 years.

I arrived in Taipei on June 2, 1926, in the midst of tropical weather quite different from that in Sendai, which is located in the northern part of mainland Japan. Contrary to my surmise, the central part of Taipei was kept very beautiful and there were many buildings surrounded by trees, such as the Government-General and the residences of high-ranking government officials. The central part of the city of Taipei was very clean, and the main streets were lined with banyan trees and other tropical plants. On seeing the Taihoku Botanical Garden, a very large garden full of rare tropical plants in full bloom, I took an instant liking to Formosa.

The population of Formosa in 1926 was about 4 million. Of this figure, almost 190,000 were Japanese, and 140,000 were of native Polynesian extraction. Most of the remaining residents were of Chinese origin, their ancestors having come from the southern parts of China (Fuchien, Shamen, or Kuangtung) more than 300 years before. Many of the Formosans spoke a Chinese dialect from the southern part of China (Fuchien or Kuangtung), but because the official language then was Japanese, most of the people spoke Japanese. Thus, one did not need to be familiar with the Chinese language in order to live or work in Formosa.

In the city of Taipei, located in the northern subtropical area, the daytime temperature exceeded 35 °C (95 °F) for more than half the year, and the humidity was very high. Naturally, there was no air conditioning. In addition, there were also daily summer thunderstorms, which sometimes surprisingly caused the tap faucets of the laboratories to spark. After Formosa became Japanese territory in 1895, the major roads in Taipei were paved, and the water supply and other public necessities were sanitized so that the central part of the city itself

Two views of the Government-General building in Taipei, Formosa, taken around 1928. Renovated after World War II, it remained in service as a government building for the Taiwanese government (Republic of China). Top: A close-up view of the building. The distinctive tower, 65 meters high, remained standing despite heavy bombardment during World War II. Bottom: The building, located near the top and center of the photo, and its surrounding environs as viewed from above. The large white building in the foreground is a museum. (Reproduced from ref. 203.)

The castle gates of Taipei around 1928. Years earlier Taipei was surrounded by a rampart with five huge gates. Although the rampart itself was replaced by a wide, tree-shaded avenue during Japanese rule, four of the gates remained. Top: The East Gate. To the right, and barely seen behind the trees, is the medical school of Taihoku Imperial University. (The Central Research Institute of the Formosan government is behind the medical school.) On the northwest corner, not shown, is the official residence of the Governor-General. Dr. Kafuku's house is on the southeast corner. Bottom: The South Gate. Across the street, the camphor oil factory and its laboratories are located on the southeast corner; the Government Monopoly Bureau on the southwest corner. (Reproduced from ref. 203.)

became quite a safe place in which to live. However, in the outskirts of the city, epidemics such as typhoid or dysentery were prevalent, and malaria was not uncommon in the suburbs.

All officers of the Formosan government were wearing official caps and uniforms, much like those of naval officers. On national holidays, even the university professors, being government employees, had to wear shoulder straps and long swords.

Research in Formosa (1926–1948)

Research Institute of the Formosan Government (1926–1929)

For the first year, I worked at the Camphor Research Laboratories of the Monopoly Bureau of the Formosan Government. For the next 2 years, I worked at the Department of Chemical Industry of the Central Research Institute of the Formosan Government-General. Both facilities were located in Taipei. At both institutes, we had to spend a considerable amount of time setting up the newly opened laboratories of the industrial organic chemistry section.

At that time in Formosa, only four chemists had university training. One of these chemists, Dr. Kinzo Kafuku, had held various positions related to chemistry. He had graduated from the Department of Chemistry of Tokyo Imperial University in 1907 and had served as a lecturer of ceramic chemistry at Tokyo College of Technology before going to Formosa in 1912. His skills and knowledge were wide and varied, and his career was very distinctive. He had served as a factory manager at a fertilizer company, as well as a mayor of Kaoshiung, in the southernmost part of Formosa. He was a person with diversified tastes and adept at both baseball and swimming. In his laboratory, he had a keen interest in devices, improvements, and inventions and had a number of patents on chemical processes and apparatus. He was also an open-hearted and humane man of extraordinary character and had very good rapport with his colleagues and junior staff. He held concurrent positions as director of both the Camphor Research Laboratories and the Central Research Institute of the Formosan Government-General and served as a professor at Taihoku Imperial University. In

Top: The east side of the two-story Central Research Institute of the Formosan government as it looked circa 1929. The Industrial Organic Chemistry Department was located on the second floor of the south side of the building (not shown). Below: The administration building of Taihoku Imperial University. It was not until several years later, in 1931, that the three three-story buildings that became Buildings I (biology and minerology), II (inorganic chemistry), and III (organic chemistry) of the faculty of science and agriculture were constructed to the right of this building. The new buildings conformed to the original in appearance. (Reproduced from ref. 203.)

1937, he returned to Japan to become a director of an industrial research laboratory of Asano Concern in Tokyo, because the president, Soichiro Asano, was an old friend of his and was eager to have him join them.

When I settled in Formosa, Dr. Kafuku was engaged in the study of essential oils of tropical plants. The research projects he assigned to me involved the mercuric acetate oxidations of linalool (at the Camphor

Dr. Kafuku in his office at the Central Research Institute of the Formosan Government-General in Taipei, 1926. He is wearing his official uniform.

Research Laboratory) and the elucidation of components of essential oils from leaves of the *taiwanhinoki,* which was one of the most important wood resources grown in Formosa (at the Central Research Institute). Dr. Kafuku emphasized that one should carry out research independently and that only work of originality was to be performed. He gave me no further instructions. This attitude provided me with a strong sense of responsibility for my own research work. Thus, my co-worker and I began collecting, with the help of the local forestry office, genuine *taiwanhinoki* (*Chamaecyparis taiwanensis* Masamune et Suzuki) leaves unmixed with the very similar *benihi* (*Chamaecyparis formosensis* Matsum.) leaves. We then obtained the essential oil by steam distillation in the laboratory. From this oil, we isolated a few interesting new terpenes[4] and sesquiterpenes.[5] However, their structures could not be elucidated because, in those days, the isolation of each component and, especially, the subsequent structural determination required a considerable amount of time.

Following the enthusiastic recommendation of Mr. Tetsu Koizumi (an anthropologist), consultant for Central Research Institute and a friend of Dr. K. Kafuku, I married Kyoko Horiuchi, Mrs. Kafuku's niece. We have one son, Shigeo, and three daughters, Takako (now Mrs. Satoru Masamune), Yoko (now married to H. Ishikura, a biochemist), and Yuriko (married to businessman K. Higashihara). In addition we now have seven grandchildren and one great-grandchild.

Unlike in the postwar era, in prewar Japan, most marriages were arranged by parents, and especially by the parents' friends. The Horiuchis, Kyoko's parents, went to Formosa with Dr. Inazo Nitobe. Kyoko's father, a graduate of the Department of Agrochemistry of Tokyo Imperial University, eventually became head of the Agricultural Station of the Formosan Government-General. The Horiuchis returned to Japan with their seven children, but after the death of Mrs. Horiuchi, Kyoko's father sent Kyoko to spend time with the Kafukus in Taiwan and asked the Kafukus to look for a suitable marriage partner. I married Kyoko after Mr. Koizumi, who was encouraging me to marry her,

Tetsuo Nozoe celebrates his 60th birthday with his family at the Japanese Inn in Spa Yugano, Izu Peninsula, May 16, 1962. From left: Mrs. Shigeo Nozoe (deceased); Yuriko (4th child, now Mrs. Higashihara); Yoko (3rd child, now Mrs. Ishikura) and her daughter, Kumiko; H. Ishikura; Tetsuo Nozoe; Kyoko Nozoe (Tetsuo's wife), holding Atsushi (Shigeo's son); and Shigeo Nozoe (2nd child).

obtained the consent of Kyoko's father and my parents (who lived in the remote cities of Morioka and Sendai in northern Japan). In those days it was even considered indecent to select one's own partner.

Taihoku Imperial University (1929–1942)

In October 1929, I was promoted to an assistant professorship in the Faculty of Science and Agriculture of the newly established Taihoku Imperial University. In the beginning, the faculty members of the Department of Chemistry consisted of two professors, two assistant professors, and four assistants. Only five students were enrolled in each undergraduate level; graduate courses were not offered. The staff were all from Japan, and most of the students were also from Japan. Very few Formosan students were enrolled in the Faculty of Science.

The availability of research funds in Formosa at that time was slightly better than that in Japan. Not only equipment but reagents, solvents, and glassware were purchased by competitive bidding, with the lowest bid being accepted. However, because of this bidding system, low-quality glassware was often delivered, and then we would have to reorder these items from Japan. This process took an additional 2 to 3 months. The situation gradually improved, especially when we built a fully equipped glass shop within our own chemistry department. Thanks to this factory, the high-quality apparatus necessary for each experiment was readily available within a few days. The head glassworker, Mr. Yoshiaki Endo, often came to our laboratory to make sure that his glassware was of good quality and appropriate for our experiments. The metal and woodworking shops in the Physics Department were also called upon to make our research run more smoothly.

As far as foreign journals were concerned, almost all of the major journals from the United States and Europe were available. However, European publications were only sent by ship via the Indian Ocean and Tokyo once every few months. Thus each publication was already several months old by the time we received it. Throughout my time at Taihoku Imperial University, I tackled three major areas of research: saponins and sapogenins (1929–1942), wool wax (1936–1942), and hinokitin and hinokitiol (1936–1942).

Saponins and Sapogenins. Gradually, I started my own research work in the midst of the construction of the new chemistry building. I worked in a poorly equipped storeroom with only the simple facilities of water, gas, and electricity. One day, Dr. Kafuku, who concurrently held the professorship at this department, told me, "When I went down

Tetsuo Nozoe with colleagues and students from the chemistry department at Taihoku Imperial University in Taipei, 1935. Front row from left: Y. Nakatsuka (assistant professor, inorganic chemistry), T. Watase (professor, analytical chemistry and a guest from Osaka University), K. Matsuno (professor, theoretical chemistry), and Tetsuo Nozoe (assistant professor, organic chemistry). Back row from left: K. Nakagawa, A. Tachiiri, T. Kinugasa (assistant, organic chemistry), K. Pan (assistant, theoretical chemistry), T. Matsumoto, S.-L. Liu, H. Matsumura, S. Ogawa, P.-Y. Yeh, Y. N. So, S. Katsura (assistant, organic chemistry), and H. Imuma (assistant, inorganic chemistry).

to a virgin forest at the southernmost end of Formosa, I found many of these curious-looking fruits lying on the ground, which looked like the legs of a checkerboard. Won't you check to see what is in the seed?" Such was the way in which one of my most important research topics (sapogenins) was initiated. After examining the plant from which this fruit came, which was identified as *Barringtonia asiatica* Kurz, a kind of mangrove, I found saponins in the seed. Hydrolysis of these saponins resulted in the formation of several kinds of sugars, tiglic acid, and two new kinds of neutral polyhydroxylic sapogenins, which I named barrigenol-A_1 and -A_2[6].

The love of plants is part of my nature, and so I found it particularly enjoyable to go on botanical trips with my students in the hills and fields nearby to collect research material. Prior to going farther into the aboriginal mountainous districts, one had to obtain special permission from the public office of the district, but I did not mind seeking this permit. Often, I went on these trips alone or with a friend, Mr.

Shinji Shono (Central Research Institute of the Government), and local natives as porters to collect medicinal plants, soap substitutes, or fish poisons containing saponins. My interest in photography commenced after I began exploring the flora in these mountainous areas.

At that time in Formosa, one usually traveled through aboriginal mountainous areas with one or two police officers for guards and one or two natives as porters. Even in the high mountains, the telephone network was well established as a security system. A police officer often doubled as an innkeeper, a teacher, or both at the elementary school for the natives. There were approximately 10 tribes native to Formosa, and the children of these tribes seemed to enjoy attending school. They were good at Japanese and especially enjoyed hunting, singing, and dancing. The people were also very friendly and honest.

In 1930, I was afflicted with acute appendicitis and had to enter a hospital. Because sulfamides or antibiotics were not known in those days, a 1-day delay often meant death. Although the appendicitis developed into peritonitis, I recovered and was able to leave the hospital in 2 weeks. During my stay in hospital, the Musha Incident, caused by the animosity between police officers and natives, occurred. In the midst of the school sports day at the elementary school of Musha (Wu-Sher), the natives suddenly attacked the Japanese people, both policemen and their families, and beheaded many of them. From the natives' viewpoint, this attack was an act of bravery. It also had religious significance. After Formosa fell to the Japanese in 1895, customs such as headhunting or head shelves became strictly illegal, and until this incident occurred, Formosa had been quite safe.

The Musha Incident reflected a return to barbarian days. The leaders of the district in which this incident occurred entrenched themselves in a deep cave within the mountains, and it took several months to suppress them. For a while thereafter, people were forbidden to go into the area. When the ban was finally lifted and access was granted from the government, I, together with my friend Shinji Shono, went for a botanical journey of 1 week to this district. From Musha, we climbed over the Noko Mountains (Nengkao-Shan, 3300 m) and traveled to the east coast. From each region, we took along two natives as porters. Because we were the first to enter the mountains after the Musha Incident, a total of 12 police officers carrying guns accompanied us over the course of the journey. The porter of Musha was a very easygoing and merry young man. When my friend light-heartedly asked him if the natives wanted to headhunt, he answered, "No, it is now prohibited," and laughed.

Eventually, we arrived at Karenko (Hualien) on the east coast of Formosa, where the Ami, a matriarchal tribe, lived. Like other natives I had encountered, they also loved to dance, and about 30 men and

Top: Aborigines hunting near Musha are carrying guns lent to them by the police. Note the stump of the hinokitiol-containing taiwanhinoki in the upper right corner. Bottom: Children of the Tsou tribe returning from farm work. The elder brother carries sweet potatoes and the younger one, taros and sugar cane. Photos taken around 1928. (Reproduced from ref. 203.)

Left: The Amis tribesmen dressed formally in native costume in Haalien, about 1928. *Right:* Gallant welcome dance of the Amis tribe. They welcomed us with this dance on our visit in 1931. (Reproduced from ref. 203.)

women dressed in gaudy costumes began to perform their gallant welcome dance for us. This journey was planned for botanical studies, but we ended up mainly taking pictures and sightseeing!

As a sideline, one of the greatest summer pastimes for baseball fans in Japan is watching the Japanese high-school baseball championships at Koshien. Several years after the Musha Incident, the Hori (Poli) team became the first representative of Formosa to participate in this event. Among the Formosan players were three natives from the mountains near Musha, and the team went on to the finals but finished as runner-up. In answer to questions posed by the reporters, one of the natives said, "We lost because we could not run. The ground was too flat, and we had to wear shoes." This reply, of course, surprised but delighted the Japanese.

With one of my assistants, Toshio Kinugasa, I continued to examine various saponin-containing plants (usually used as soap substitutes or fish poisons), such as *Barringtonia racemosa* Blume[7] and *Schima kankaoensis* Hay.[8] Glycosides of neutral polyhydroxytriterpenoids, esterified with tiglic or angelic acid, were found to be widely distributed throughout nature. In those days, studies on sapogenins were being conducted actively and were one of the main topics in the field of organic natural products chemistry, particularly in Europe and Japan. However, because of the difficulty in separating similar neutral sapogenins and determining their molecular formulas, I decided to postpone the structural determination of the complex polyhydroxysapogenins.

I began structural studies on oleanolic acid and hederagenin. These compounds were more readily available and were known as representative pentacyclic triterpenoids.[9] The reported structural formulae for these substances were inconsistent, despite the very active involvement of researchers from various countries. Sapogenins of this type were generally thought to be pentacyclic triterpenoids of the picene skeleton, on the basis of the pioneering work of Ruzicka.[10] Selenium dehydrogenation of pentacyclic sapogenins of β-amyrin type yielded 1,8-dimethylpicene (5), as well as benzene and naphthalene homologues. Various chemical structures regarding pentacyclic sapogenins were proposed; however, none of them satisfied the experimental results.

During that time (1932–1939), the study of sapogenins was popular, but UV spectroscopy was only in limited use. In those days, measurement of UV–visible absorption spectra was carried out by allowing extremely complex iron arc lines (2300–5700 Å) to pass through a sample solution placed in a Baly tube while varying the path length. The transmitted spectra were printed on a special photographic plate, and any changes in the intensities or disappearance of the absorption lines at each wavelength were closely examined by using a telescope. Then, after some calculation, absorption curves (log 1/transmittance versus wavelength) were obtained. Because such a procedure was extremely laborious and time-consuming, UV spectroscopy was used mainly in the study of dyestuffs and natural pigments, and in the field of physical organic chemistry to see the relationship between conjugated double bonds and the absorption spectra. The technique was rarely used for the structural study of colorless natural products at that time.

However, I tried to use UV absorption spectroscopy as much as possible for the study of the structures of reaction products. By oxidation of these widely distributed sapogenins with various reagents (CrO_3, $KMnO_4$, SeO_2, O_3, C_6H_5COOOH, and H_2O_2), we obtained, in addition to several degradation products,[11] ketolactones **6** (end absorption), as well as isoketo acids **7** (λ_{max} 245 nm), keto acids **8a** and **8b** (λ_{max} 245 nm), dehydro acids **9** (λ_{max} 245 nm), and isodehydro acids **10** (λ_{max} 290 nm), as shown in Chart I. Tetradehydro (**11**) (λ_{max} 310 nm) and bromotetradehydro derivatives (λ_{max} 310 nm)[11,12] were produced in high yield by the action of bromine in acetic acid or ethanol at low temperatures. Because the position of the methyl groups in the previously presented structures for these sapogenins appeared to be incorrect in view of our UV spectral results, especially the formation of conjugated trienes (**11**), we proposed the structures **12a** (R = CH_3) for oleanolic acid and **12b** (R = CH_2OH) for hederagenin, which are in agreement with the isoprene rule, in 1937 in short communications.[12] We reported the experimental details the following year.[13] On later examination of the literature, I found that soon after the publication of our independent study (in 1937), Ruzicka and others had also begun to use UV spectroscopic measurements in the structural study of sapogenins and lanosterol and had obtained similar results (*vide infra, see* Overseas Research on Triterpenoids and Wool Wax).

To derive other natural products from the two sapogenins as the acetates, **13a** and **13b**, we hydrogenated the acid chlorides **14a** and **14b** in the presence of palladium on carbon. To our surprise, the reaction resulted in the formation of formaldehyde and the conjugated dienes **15a** and **15b** ($\lambda_{max}^{CH_3OH}$ 245 nm) in high yields.[13] We could not obtain the aldehyde derivative from **14** by the usual Rosemund reduction, whereas the Bouveault–Blanc reduction of methyl esters of **13a** and **13b** yielded erythrodiol and hederatriol (**16a** and **16b**, respectively), and the acetylhederagenin **13b** was successfully converted into naturally occurring gyposogenins via hederagic acid.[14] We discovered later that Ruzicka et al.[15] had also succeeded at the same time in correlating oleanolic acid with erythrodiol via a corresponding aldehyde obtained by the Rosemund reduction. We reported a part of these degradation studies in short communications[11,12] and in a national meeting.[13,14]

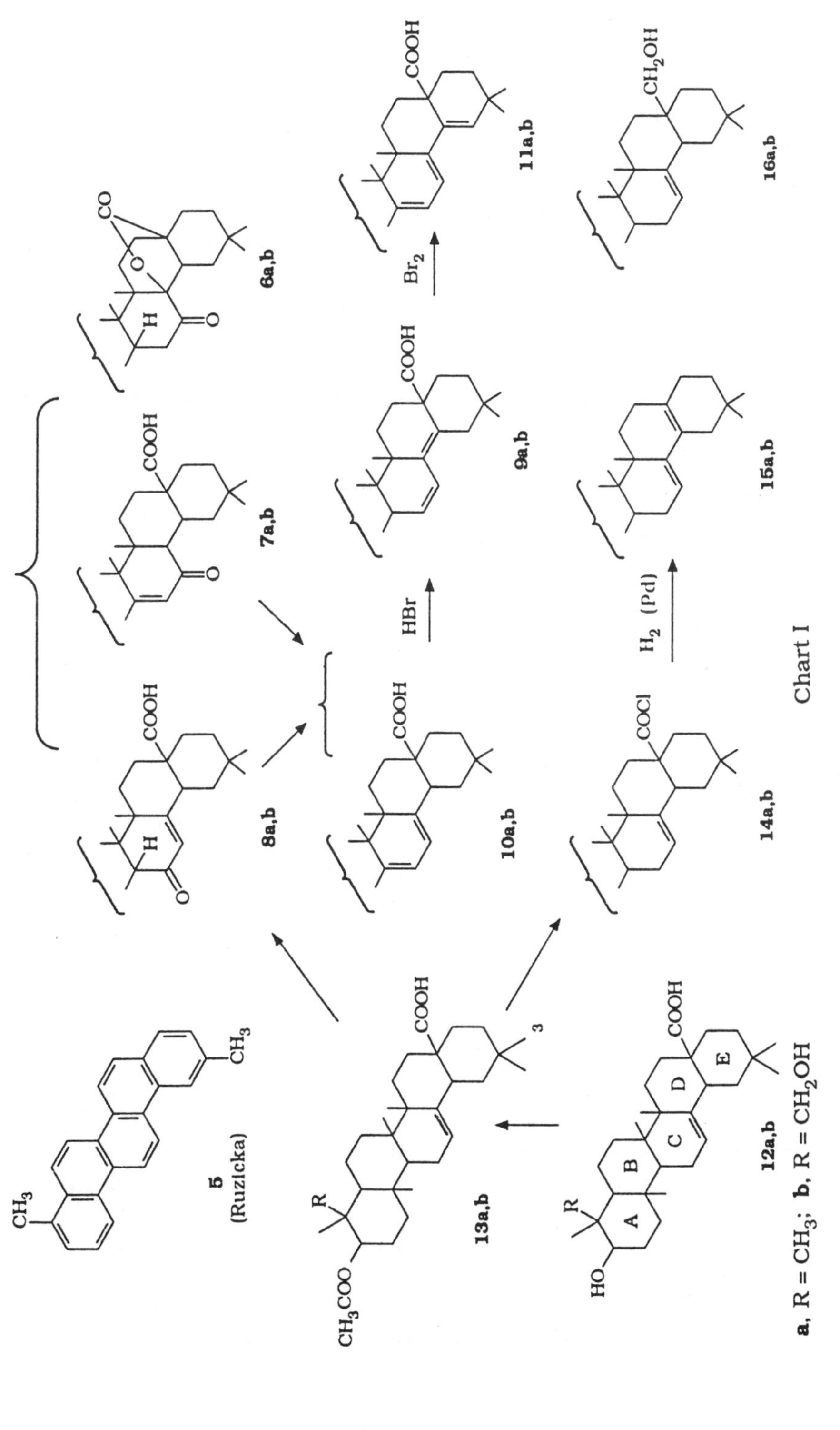
5
(Ruzicka)
8a,b
7a,b
6a,b
10a,b
9a,b
11a,b
13a,b
14a,b
15a,b
16a,b
12a,b
HBr
Br_2
H_2 (Pd)
COOH
COCl
CH_2OH
CH_3COO
HO
R
A
B
C
D
E
a, R = CH_3; b, R = CH_2OH

Chart I

After 1942, further detailed structural investigation had to cease because of the intensifying war.

Along with the structural elucidation of sapogenins, I carried out a systematic study of the color reaction of the sugar part of saponins with various phenols and arylamines by using a visible spectrophotometric method. In those days, we generally used color reactions for the identification of the sugar type.[16] After heating a sugar mixture with various phenols or arylamines in concentrated hydrochloric acid and then extracting the colored reaction mixture with isoamyl alcohol, one could identify the sugar type either by observing the color change of the organic layer with the naked eye or a Dubosq colorimeter or by observing the change of the absorbed wavelength by using a very simple spectroscope. However, I had not been satisfied with this method, which concentrated primarily on the differences of the exhibited color shades and not on the structures of the pigments themselves. For this reason, by observing the time-dependent changes of the visible absorptions of the isoamyl alcohol layers, I examined various known color reactions of several monosaccharides and furfural derivatives that were expected to be produced from monosaccharides by heating the reaction mixture with hydrogen chloride.

Because it was impossible to observe time-dependent changes of visible absorption spectra by traditional methods, I managed to borrow a very expensive Nutting photometer, which was kept in a glass show cabinet in Professor Kichizo Matsuno's office. At first the professor would not agree to my request to borrow the photometer, but eventually he relented. I carried out the time-dependent measurement by hand. The wavelength was varied by rotating a drum with one hand; the absorptions were measured with the naked eye and quickly written down with the other hand. Thus, I was able to obtain[17] the absorption curves of various pigments derived from the species that had been produced in the reactants, although the task of repeating such measurements in a small, dark room at 40 °C with almost 100% relative humidity in this subtropical area was extremely difficult. Examples of the curves obtained in the Bial reactions of arabinose, rhamnose, and fructose[17] are shown in Figure 1.

These absorption curves (Figure 1), which had been published more than 50 years ago, are similar to the more-accurate figures currently obtained by modern automatic spectrophotometers. However, because the study of sugar color reactions was not my main area of research at the time, I decided not to become too involved in this problem. It was not easy in those days to separate a mixture of water-soluble pigments. Nevertheless, essentially the same methodology as the one described here has now been developed with modern, automatic time-dependent high-performance liquid chromatographs

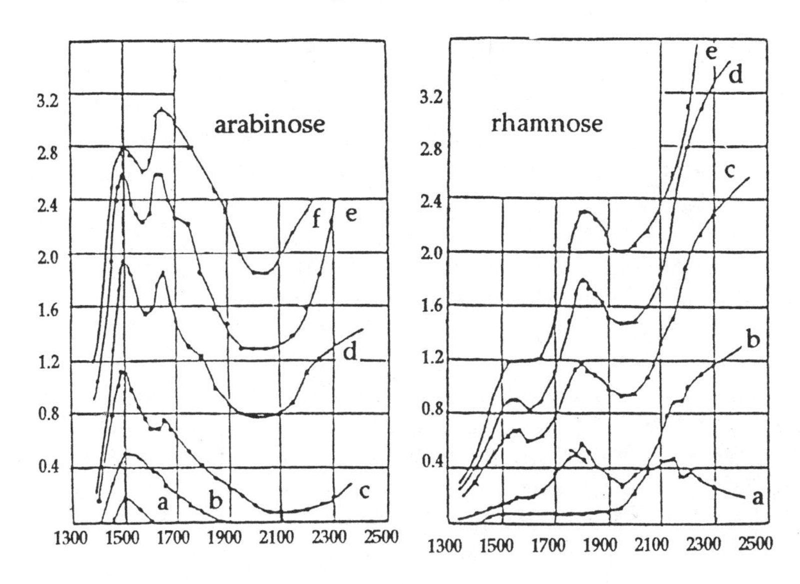

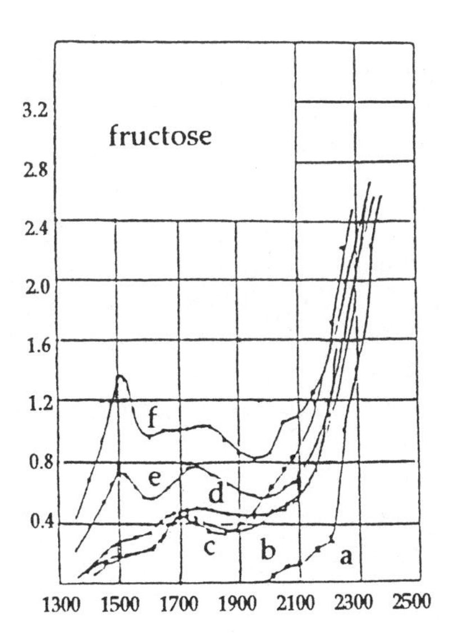

Figure 1. Time dependence of the absorbance curves of the Bial reaction of monosaccharides. The reaction times in minutes for the different curves are as follows: arabinose, a = 2, b = 3, c = 5, d = 8, e = 15, and f = 30; rhamnose, a = 1, b = 2, c = 3, d = 5, and e = 8; fructose, a = 0.5, b = 1, c = 1.5, d = 2, e = 3, and f = 5. (Reproduced with permission from ref. 17. Copyright 1936 Chemical Society of Japan.)

with UV–visible detectors, which enable one to monitor even highly competitive, complex reactions in solution from the beginning to the end. This capability was exemplified by our recent studies on troponoid and azulenoid compounds (*see* Mechanism of Our Azulene Synthesis).

It was at this time that Professor Majima had been appointed Dean of the Faculty of Science of Osaka Imperial University, established in 1931. In 1943 he was selected its President. From 1935, Osaka University offered a D.Sc. degree by way of presentation of a dissertation. On advice from Professor Majima, I summarized the research that I had studied on organic natural products and submitted my results to the university administration through Professor Munio Kotake, an elder graduate of Tohoku Imperial University. In 1936, I became the first recipient of a D.Sc. degree from Osaka University. The following year, upon Dr. Kafuku's return to Japan, I was appointed a full professor at Taihoku Imperial University.

In March 1938, I invited Professor Majima to Formosa. During his 2-week stay, including a 1-week journey by boat, I was with him for 6 days while we traveled to the southern end of Formosa. During this time, we had the opportunity to talk not only about our research but also about various other topics, and for the first time, I learned why he had advised me to go to Formosa immediately after graduation, without letting me enter the postgraduate course. He and Dr. Kafuku had determined that it was in my best interest to accept the post in Formosa. Because I was to be invited as an assistant professor initially, they thought that working in Formosa would give me much more freedom to do my own research than would staying in Japan as a postgraduate student. Together with Dr. Kafuku, Professor Majima had planned to make me a full professor in Taihoku Imperial University after I obtained my doctorate.

Furthermore, in Formosa, where chemists were few and research facilities were scarce, I had to rely on my own abilities for all aspects of my work: preparation of the laboratories and facilities, selection of a research theme, and choice of a method by which we proceeded with the research. Neither Professor Majima nor Dr. Kafuku had given me detailed instructions regarding my research; they thought I would benefit most if I was given the freedom to use my own abilities. I was very moved by their thoughtful consideration. If I had been given explicit, detailed instructions by the professors day in and day out, like many postgraduates before and even now, I would not have been able to achieve the aforementioned results in Formosa.

Wool Wax. The major difference between wool wax (or wool fat) and common fat or oil is that wool wax is known to consist of an equal quantity of acids of unknown structure and "unsaponifiable matter".[18]

Dr. Kafuku in front of his house in Tokyo, 1940, after his return to Japan.

The staff, students, and technicians of the chemistry department of Taihoku Imperial University in Taipei, 1938. Front row, second from left is Tetsuo Nozoe (professor, organic chemistry). Behind him are Y. Nakatsuka (professor, inorganic chemistry), K. Ochiai (professor, theoretical chemistry), and N. Ikegami. Second row from left: H. Imuma (assistant professor, inorganic chemistry), M. Kainosho, L.-S. Chen, T. Kinugasa (assistant professor, organic chemistry), Shigeo Katsura (assistant, organic chemistry), S.-L. Liu, H. Matsumura, and I. Tominaga. Those in the third row and at the ends of the first and second rows are students and technicians.

The unsaponifiable matter was presumed to contain a large amount of lanosterol and agnosterol, in addition to wax alcohol (ceryl alcohol) and cholesterol. Lanosterol was believed to be the same compound as the triterpenoid cryptosterol found in yeast. Windaus[19] and Wieland[20] were reported to withdraw from their research on lanosterol, because it was not considered a real sterol. On the contrary, I was particularly interested in the fact that an animal product like wool wax contained triterpenoids that are usually found in plant products. In 1936, with my assistant Shigeo Katsura, I decided to study the structure of lanosterol and compare it with pentacyclic sapogenins.

I obtained authentic wool wax from a former classmate at Tohoku Imperial University, Dr. Yasota Kawakami, who was then the manager of research and development at the Kao Soap Company in Tokyo. In Japan, persons of academic rank at national universities are forbidden to accept positions as consultants of industrial companies and accept honoraria. However, a company is allowed to support research by providing materials or funds. Hence, in this manner, Dr. Kawakami was able to assist me privately.

I hydrolyzed about 500 g of wool wax with ethanolic potassium hydroxide and then carefully separated the neutral portion from the acidic portion by extracting the product mixture with either diethyl ether or petroleum ether. This saponified material had a strong emulsifying property. Systematic fractional extraction (the prototype of countercurrent distribution) with several large separatory funnels and a low-boiling-point solvent was very difficult because of the high temperatures in the laboratory. Because no freezer was available, we cooled the material with ice and conducted the experiment as far away from any flame as possible. We found through the available literature that both the neutral and acidic components, which we had painstakingly separated, consisted of a mixture of unknown compounds. Because the acidic component was speculated to contain terpenoid acids, I decided to study all of the components of wool wax, in addition to the sapogenins.

Although the presence of branched-chain or cyclic fatty acids, in addition to the normal straight-chain fatty acids with even-numbered carbon atoms and α-hydroxypalmitic acid (lanopalmic acid), in the acidic fraction of wool wax had been speculated,[21,22] the structures of these products were, as yet, unknown. A branched-chain fatty acid, tuberculostearic acid, found in the tubercle bacillus, was the only known compound contained in natural waxes, besides hydrocarpic and chaulmoogric acids that have the cyclopentane ring and are known to exist as triglycerides in chaulmoogra oil. Thus, as a result of my unusual curiosity in organic chemistry to inquire about the unknown, I, along with my student, Sheng-Lieh Liu, started to investigate the acidic portion of wool wax.

The whole fatty acids obtained were converted to methyl esters, from which hydroxy acid methyl esters were completely removed by repeated large-scale alumina column chromatography. The crude mixture of nonhydroxylic esters was separated into pure fractions first by systematic fractional vacuum distillation and then by extremely careful fractional distillation (drop by drop). During this procedure, we checked not only the boiling point but also the optical rotation by using a specially prepared microobservation tube. We finally succeeded[23] in obtaining pure fractions of the optically inactive C_{12}–C_{20} fatty acid esters with an even number of carbon atoms (**17**) and the dextrorotatory C_{11}–C_{21} fatty acid esters with an odd number of carbon atoms (**18**) (Figure 2).

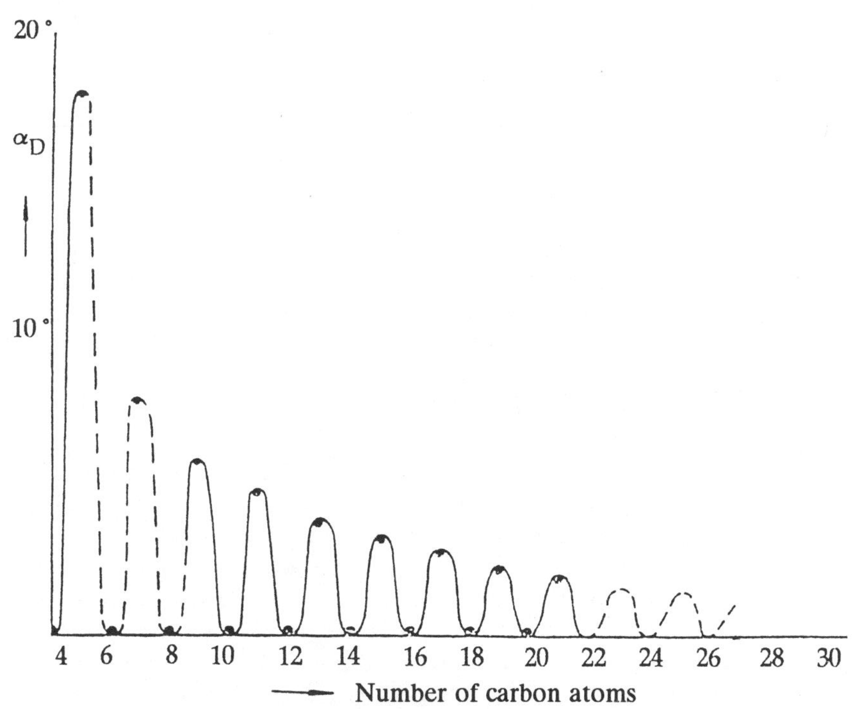

Figure 2. Optical rotation, $[\alpha]_D$ (in a 1-dm microtube), of fractions obtained by careful fractional distillation of the methyl ester of nonhydroxylated wool wax acids. The y axis gives the number of carbon atoms in the wool wax acids. (Reproduced with permission from ref. 25. Copyright 1952 National Taiwan University.)

Hydrolysis of the optically inactive esters reproduced the free acids, with melting points almost identical to those of the naturally occurring straight-chain fatty acids with the same number of carbon atoms. However, these two corresponding even-numbered-carbon fatty acids (from wool wax and normal acid), which had almost identical melting points, showed surprisingly large depressions in their mixed melting points. Without complete purification, wool wax acids do not crystallize well. The purified, even-numbered-carbon fatty acids in our laboratory always produced beautiful plates, whereas the odd-numbered-carbon fatty acids produced needlelike crystals. The structures of the abnormal fatty acids were established by degradation with chromic acid to produce dicarboxylic acids with odd-numbered-carbon atoms (**19**) and acetone (from **17**) or methyl ethyl ketone (from **18**).[23]

$$(CH_3)_2CH-(CH_2)_{2n}-COOH \quad \mathbf{17} \longrightarrow CH_3COCH_3 + COOH-(CH_2)_{2n-1}-COOH$$

$$CH_3CH_2(CH_3)\overset{*}{C}H-(CH_2)_{2n}-COOH \quad \mathbf{18} \longrightarrow CH_3CH_2COCH_3 + COOH-(CH_2)_{2n-1}-COOH$$

$$COOH-(CH_2)_{2n-1}-COOH \quad \mathbf{19}$$

The water-soluble, low-molecular-weight carboxylic acids obtained after the hydrolysis of wool wax, butyric and isovaleric acids, had been reported earlier. However, we did find isobutyric acid and a new dextrorotatory valeric acid.[24] We also found no normal fatty acids with less than 20 carbons present in the non-hydroxylic acid part extracted from the wool. The structures and properties of our newly isolated wool wax fatty acids are shown in Table I.[23,25]

The separation of the high-molecular-weight portion (C_{22} or more) of wool wax fatty acids was very difficult. We used repeated systematic fractional recrystallization of various functional derivatives (such as amides, *p*-toluidides, or 2,4,6-tribromoanilides), but we could not complete the separation of the fatty acids. Nevertheless, we confirmed the presence of a normal fatty acid ($C_{26}H_{52}O_2$) and its branched-chain isomer, as well as dotriacontanoic acid ($C_{32}H_{64}O_2$).[24,25] We found that the "lano acids" (C_{14}–C_{20}) obtained by Kuwata[21] were very impure and confirmed that these acids were a mixture of the two series of acids, **17** and **18**. Thus we named **17** and **18** the lano (lanolinic) and agno (agnolic) acid series, respectively.[23]

The ester portion of the hydroxy acids was isolated first by alumina chromatography and then by fractional distillation. The free acids were obtained by alkaline hydrolysis. In addition to the previ-

Table I. Physical Properties of Lanolinic and Agnolic Acids from Wool Wax

Acid	*Formula*	*Structure*[a]	*Melting Point (°C)*		*Optical Rotation of Methyl Ester*[b]
			Acid	*Amide*	
Isobutyric	$C_4H_8O_2$	17 (n = 0)	liq.	75.5[c]	±0.00
Agnovaleric	$C_5H_{10}O_2$	18 (n = 0)	liq.	52.[c]	+16.56
Agnocarpic	$C_{11}H_{22}O_2$	18 (n = 3)	liq.		+4.90
Lanolauric	$C_{12}H_{24}O_2$	17 (n = 4)	42.5		+0.48
Agnolauric	$C_{13}H_{26}O_2$	18 (n = 4)	liq.	94	+4.08
Lanomyristic	$C_{14}H_{28}O_2$	17 (n = 5)	55.5	107	+0.25
Agnomyristic	$C_{15}H_{30}O_2$	18 (n = 5)	28.0	90	+3.52
Lanopalmitic	$C_{16}H_{32}O_2$	17 (n = 6)	62.0	101	±0.00
Agnopalmitic	$C_{17}H_{34}O_2$	18 (n = 6)	41	92	+3.22
Lanostearic	$C_{18}H_{36}O_2$	17 (n = 7)	70	108	±0.00
Agnostearic	$C_{19}H_{38}O_2$	18 (n = 7)	51	95	+2.86
Lanoarachidic	$C_{20}H_{40}O_2$	17 (n = 8)	76.0	103	±0.00
Agnoarachidic	$C_{21}H_{42}O_2$	18 (n = 8)	58.0	98	+2.18

SOURCE: Reprinted with permission from ref. 25. Copyright 1952 National Taiwan University.

[a]Number (n) of methylene groups for lano acid series 17 and agno acid series 18 are shown here.

[b] All values are degrees. Measurements were made in 1-dm microtubes.

[c] The melting point of the *p*-bromophenacyl ester, instead of the amide, is given.

ously known α-hydroxypalmitic acid (or lanopalmic acid),[21] α-hydroxylauric acid, α-hydroxymyristic acid, and α-hydroxylanostearic acid were obtained as optically active levorotatory forms and racemic forms.[26] The structure of a high-molecular-weight lanoceric acid, obtained as a precipitate during the washing of wool with soda, turned out to be ω-hydroxydotriacontanoic acid ($C_{32}H_{64}O_3$).[26]

It took nearly 6 months to separate all of these acids. In retrospect, I feel that we performed the separation quite well, considering that we achieved the task without the modern equipment available today. I was really surprised to find a complex specificity in the acidic portion of wool wax, and I was anxious to know why it contained such complex and unusual components. Therefore, we collected skin wax from horse, cow, dog, and human hair, as well as face skin lipid (courtesy of the Kao Soap Company in Tokyo), but we were never able to find any branched fatty acids in these waxes.

As for the neutral fraction of wool wax, separations of "ceryl alcohol," cholesterol, and "isocholesterol" proved to be very difficult. We performed the preliminary separation of these neutral constituents by using a homemade molecular distillation apparatus to give three fractions: low-molecular-weight components, cholesterol, and isocholesterol. These crude fractions were purified by alumina chromatography and recrystallization of their functional derivatives, such as their acetates, benzoates, or *p*-bromobenzoates.

Branched-chain alcohols and normal alcohols were found in the neutral fraction, as in the acidic fraction. Although the complete separation and confirmation of these complicated constituents were not accomplished, we isolated two types of glycols: α-lanocerol ($C_{22}H_{46}O_2$, mp 74–75 °C) and β-lanocerol ($C_{24}H_{50}O_2$, mp 85–86 °C). From the formation of the acetone condensation product and the results of Criegee degradation, we confirmed that β-lanocerol had the structure **20**.[27]

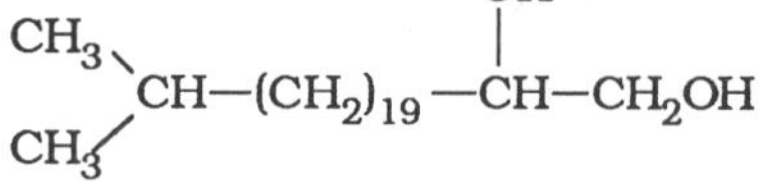

$$(CH_3)_2CH{-}(CH_2)_{19}{-}CH(OH){-}CH_2OH$$

20

Although isocholesterol was thought to consist of lanosterol and a small amount of agnosterol and was considered a pentacyclic triterpenoid, the structure remained completely unknown.[18–21,28] Like Windaus,[19] we recrystallized isocholesterol as an acetate and were able to separate α- and β-lanosterol and α- and β-agnosterol.[29] However, depending on the source materials, only α- and β-lanosterol were obtained, and no agnosterols were present. We proposed that agnosterol was a secondary product resulting from the autoxidation of lanosterol.[29]

The oxidation of lanosterol acetate after catalytic reduction with chromic oxide in acetic acid gave a deep-yellow diketone, whereas oxidation without prior catalytic reduction yielded the diketone and acetone, as well as a yellowish diketo acid.[29] The hydrolysis of these two (neutral and acidic) diketo acetates yielded a deep-yellow diketone ($C_{30}H_{40}O_2$, which we named lanochromin) and a diketo acid ($C_{27}H_{40}O_4$, which we named lanochromic acid). To lanochromin and lanochromic acid, we assigned the partial structures **22a** and **22b**, respectively, on the basis of the UV spectra (λ_{max} 274 nm), and to both lanosterol and dihydrolanosterol (α and β), we assigned the partial structures **21a** and **21b**, respectively.[29] A part of these investigations were presented at an annual meeting of the Chemical Society of Japan.[29,30]

In 1940, I collected all the experimental data I had obtained until then and sent them to my friend, Dr. Kawakami of the Kao Soap Company, to see if the reports would be accepted by the *Journal of the Chemical Society of Japan* as full papers. Some of the reports were published in the *Journal of the Chemical Society of Japan* as short communications. It was impossible to publish long papers in a journal during the war. As a result, Dr. Kawakami added some practical applications to the data and

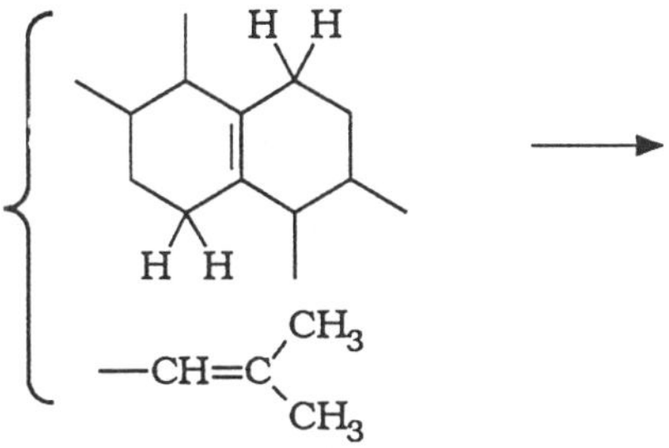

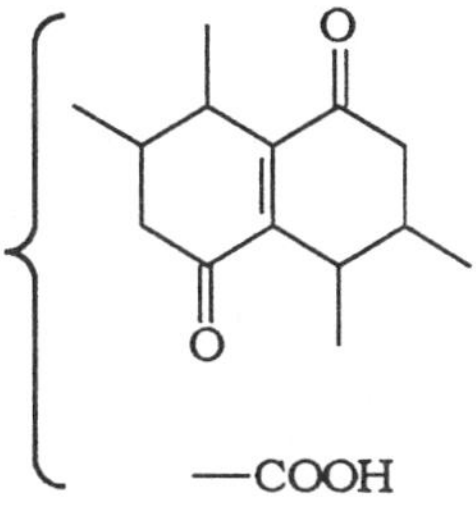

—CH=C(CH_3)$_2$ —COOH

21a **22a**

2H

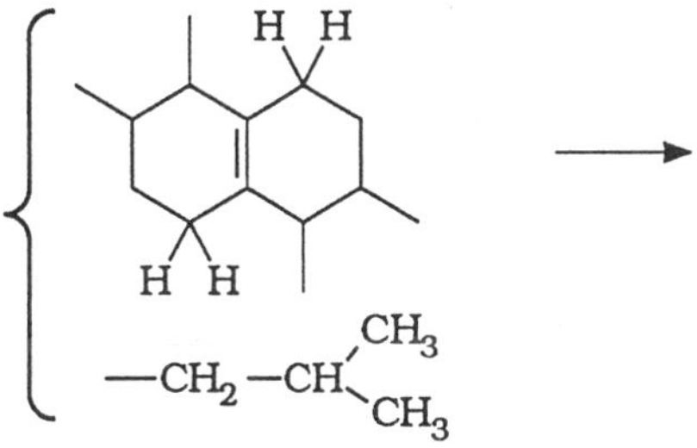

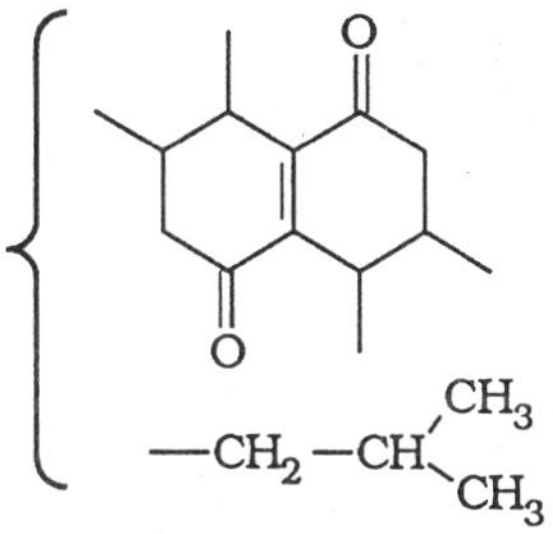

—CH_2—CH(CH_3)$_2$ —CH_2—CH(CH_3)$_2$

21b **22b**

applied for 10 patents, mostly under my name. The applications were accepted, and the patents were made public in 1941–1942. Among these patents were the aforementioned structural determinations and physical properties of the new compounds.

The continuation of further studies on these topics also became impossible because of the wartime difficulties. I still had a special interest in discovering the meaning of the existence of the unusual waxes in wool. Some questions I wanted to answer were (1) Did the various types of acids (including normal and branched-chain fatty acids and their hydroxy derivatives) and alcohols (including various types of fatty alcohols and polycyclic alcohols, such as cholesterol and lanosterol) that constitute wool wax form ester bonds with each other at random? (2) Did any of the special esters or ketones (lanochromin and other polycyclic ketones found in wool) have any physiological or physical meaning?

To answer these questions, I soaked the wool in petroleum ether repeatedly at room temperature and examined the optical rotation of the lipid obtained from each extract. The observed optical rotation, $[\alpha]_D$, rapidly dropped from +57° (first extract), to +35° (third extract), and finally to −20° (sixth extract). The resulting large dextro- and levorotation were obviously based on lanosterol ($[\alpha]_D$ +61°) and cholesterol ($[\alpha_D$ −30°), respectively. Agnolic acids (18) and hydroxy acids are dextro- and levorotatory, respectively, but the $[\alpha]_D$s of both acids were small, so I assumed that the large dextro- and levorotation of the aforementioned wax fractions were caused by lanosterol (or agnosterol) and cholesterol, respectively. On the basis of these findings, grease from the sebaceous glands of sheep should have consisted mainly of lanosterol wax, and the final extract should have contained cholesterol wax, which existed in the cells of surface skin. This point and the physiological or physical meanings of the aforementioned components of wool wax had not yet been elucidated. At that time, it had not yet been clarified that lanosterol was an important precursor of cholesterol during biosynthesis. Unfortunately, beginning in 1942, we had to completely terminate our research because of the war effort. The precious experimental materials, mammalian lipids, were lost during our evacuation.

Hinokitin and Hinokitiol. During this time, I was also involved in a small research project at Taihoku Imperial University. While I was reviewing my previous study of the essential oils of the *taiwanhinoki*, I became interested in the components of the essential oils of three important conifers: *taiwanhinoki*, *benihi*, and Japanese *hinoki* (*Chamaecyparis obtusa* Sieb. et Zucc.), all of which looked very similar to us. However, these conifers are extremely different in their constituents, which vary depending on the species and the part (leaf or heartwood) of the plant. I was particularly attracted to the acidic substances that were present in minute quantities in these essential oils.

It had already been reported[31] that the acidic portion of the heartwood of the *taiwanhinoki* contained a small quantity of a supposedly phenolic substance ($C_{10}H_{12}O_2$), whereas a dark-red wood pigment "hinokitin," erroneously assigned the formula $C_{30}H_{34}O_{10}$, was isolated by Hirao[32] from the same oil. Hirao speculated that the pigment was derived from an unidentified acid ($C_{10}H_{16}O_2$) by oxidation. Another supposedly phenolic substance ($C_{10}H_{12}O_2$), which had been reported by Kawamura,[33] was isolated from the acidic portion of the wood oil of *hiba* (*Thujopsis dolabrata* Sieb. et Zucc.), which also belongs to the same class, *Cuppressaceae*, that grows in northern Japan. This substance was strongly resistant to wood-decaying fungi. A common characteristic of all these acidic substances was the red coloration with ferric chloride. Therefore, I speculated that a certain common substance may be present in the acidic portions of those wood oils.

On shaking an ethereal solution of hinokitin with an aqueous alkaline solution, I obtained a gelatinous precipitate of ferric hydroxide and an enolic compound having a molecular formula of $C_{10}H_{12}O_2$ as the alkaline salt. I named the compound "hinokitiol".[34] I also confirmed that hinokitiol was present in *hiba* oil but not in the Japanese *hinoki* or *benihi* oil. My experiment proved that hinokitin, which had been considered a natural pigment, is an iron complex ($C_{30}H_{33}O_6Fe$) of hinokitiol instead. The percentages of C and H in Hirao's formula, which were obtained by elemental analysis, were also correct for the iron complex. It is understandable then not to have noticed the presence of Fe in such a natural compound that was sublimable in vacuo and soluble in organic solvents (diethyl ether or chloroform).

On the basis of experimental facts that hinokitiol was a monobasic acid that gave acetone by oxidation with various reagents and readily formed neutral metallic complexes with almost any metallic salt, I first proposed formula **23** for hinokitiol. However, I soon realized that the composition and physical properties of the metal complexes of hinokitiol were entirely different from those of *o*-hydroxyacetophenone derivatives. I then considered the five-membered β-diketone (**24**) and the six-membered α-diketone (**25**), as well as a formula containing an isopropenyl group (**26**). However, none of these structures seemed to be consistent with the properties of hinokitiol. Although the enol form (**27**) of the seven-membered α-diketone was the last remaining possibility, the common knowledge at that time that no such compound could exist in nature in a stable form led me to abandon the formula for a while.

In 1936, a special issue of the *Bulletin of the Chemical Society of Japan* had been planned in celebration of Professor Riko Majima's 60th

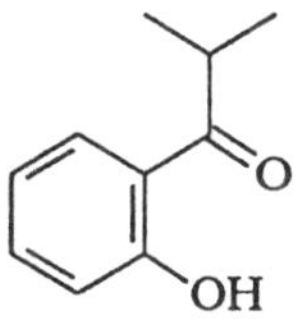

23 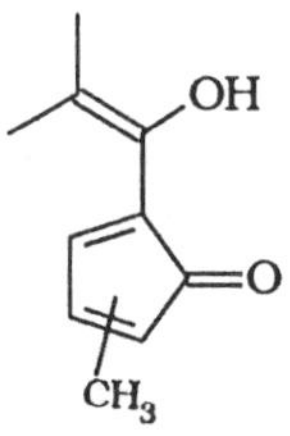 **24** 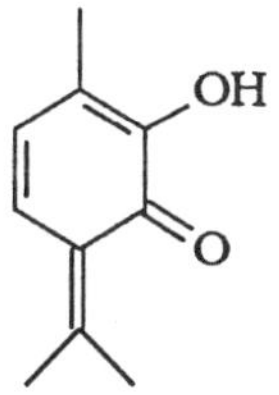 **25**

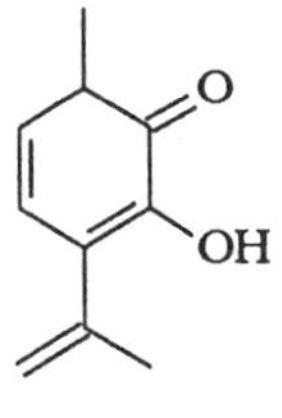

26 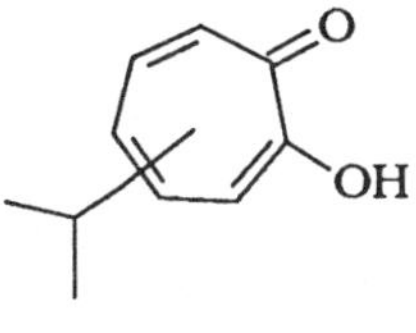 **27**

birthday. Because he had shown a continued interest in this research problem, I decided to dedicate a paper to my former supervisor for the celebration. Being unable to solve the proper structure, I simply described hinokitiol as a monocyclic α- or β-enolone having an isopropyl group[34] and that readily formed complexes with virtually all metals. This report happened to be the first of over 500 publications on our troponoid studies, which have continued to this day, some 50 years later. However, for lack of hinokitiol supplies and because of other concurrent research projects, such as the work on triterpenoids and wool wax, I was forced to interrupt the hinokitiol study for a few years.

A little later, in 1938, Professor Shigehiro Katsura, who had been newly appointed to the Faculty of Medicine of Taihoku Imperial University, came to me and asked for my cooperation in his research project on the chemotherapy of tuberculosis by the use of fatty acids and related materials. I gave him some wool fatty acids, *l*-rhodinic acid (with a terpenoid structure) from *taiwanhinoki* oil, and a small quantity of hinokitiol. Half a year later, Professor Katsura told me that both *l*-rhodinic acid and hinokitiol exhibited significant antibacterial activities against a tubercle bacillus and other bacteria. He then earnestly requested that I acquire larger quantities of these materials. Thus, I visited the Taipei factory of Takasago Perfumery Company, Ltd., and asked for their cooperation. The chief executive of the company, Dr. Teikichi Hiraizumi, readily agreed to my request. They kindly collected stumps of *taiwanhinoki* left in the mountains and obtained the hinoki oil, from which *l*-rhodinic acid and hinokitiol were separated by my former student Mr. Masayasu Kainosho. Thus, the Takasago Perfumery obliged us by donating a considerable amount of these samples for medical research.

About 10 g of hinokitiol was also given to me for structural study, and so I resumed the elucidation of the hinokitiol structure with my assistant, Shigeo Katsura. We conducted a series of degradative studies; catalytic reduction of hinokitiol gave the saturated diol **28** ($C_{10}H_{20}O_2$) as the main product. The diol was then subjected to Criegee degradation and H_2O_2 oxidation. The product was the dicarboxylic acid **29** ($C_{10}H_{18}O_4$). Because pyrolysis of **29** yielded the cyclic ketone **30** ($C_9H_{16}O$), we considered that hinokitiol must be a monocyclic α-enolone with a six- or seven-membered ring. Today, the structure could have been determined very easily by using IR or NMR spectra. In those days, however, structural determinations depended almost exclusively on chemical reactions. Because hinokitiol is not converted into a catechol derivative by isomerization with acid or alkali, both the six-membered-ring formulas (**25** and **26**) had to be eliminated. Combinations of these results led me to reconsider the seven-membered structure, **27**, for hinokitiol.

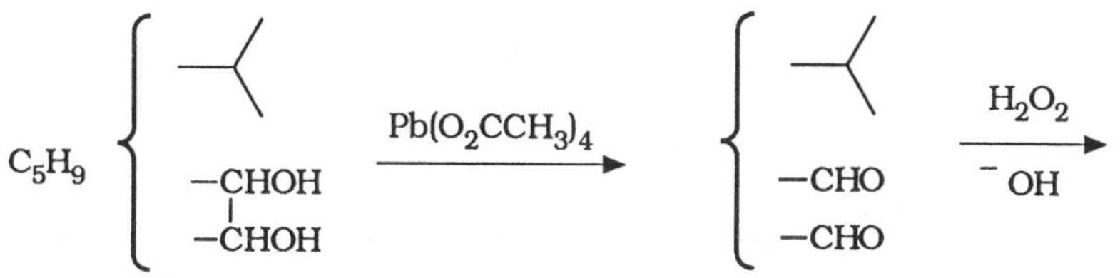

C_5H_9 —CHOH —CHOH $\xrightarrow{Pb(O_2CCH_3)_4}$ —CHO —CHO $\xrightarrow[^-OH]{H_2O_2}$

28

—COOH —COOH $\xrightarrow{\Delta}$ =O

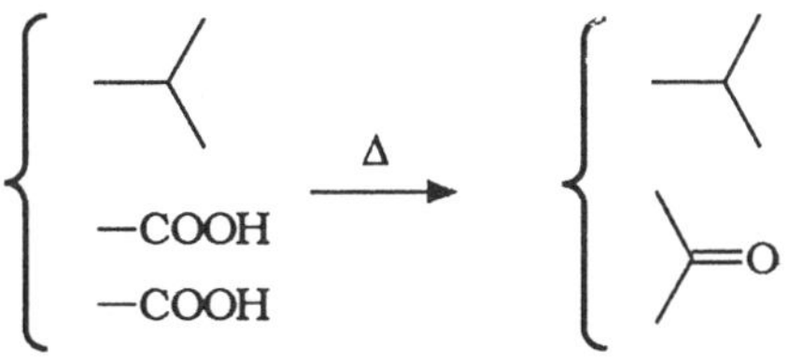

29 **30**

Then, to determine the position of the side chain, we conducted various oxidative studies on hinokitiol, but we encountered many difficulties. For example, strong oxidizing agents caused a complete breakdown of the skeleton, whereas oxidation under milder conditions, with $KMnO_4$ or CrO_3, resulted in the formation of metal-containing resinous products. Oxidation with alkaline hydrogen peroxide, however, yielded compound B (β-isopropyllevulinic acid, **31**), in addition to compound A (a dicarboxylic acid, $C_9H_{12}O_4$), the structure of which remained unclarified until then. Because the oxidation of hinokitiol yielded a considerably high amount of the saturated keto acid $C_8H_{14}O_3$, we presented,[35] in 1940, a tentative structure, **32**,[36] for hinokitiol that had one less double bond and thus had the molecular formula $C_{10}H_{14}O_2$, by strong request of Shigeo Katsura, who studied H_2O_2 oxidation of hinokitiol.

After the annual meeting of the Pharmaceutical Society in Tokyo, Shigeo Katsura joined the Takasago Perfumery Company as a member of the research staff and remained in Japan. I returned to Formosa and continued the structural study alone. I soon reconfirmed that the molecular formula of hinokitiol should have been $C_{10}H_{12}O_2$, on the basis of elemental analyses of the monoacetate and the monomethyl

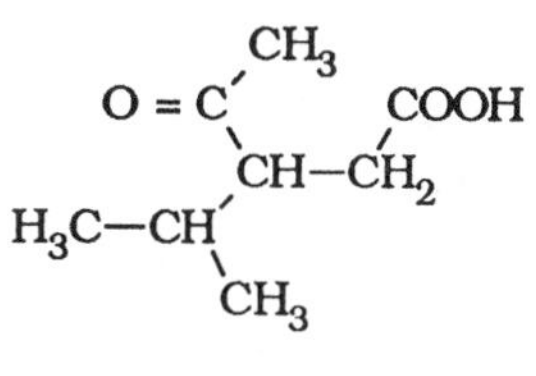

31 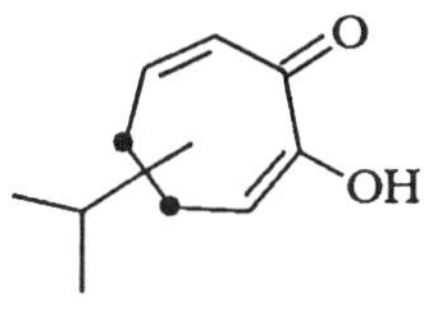 **32**

ether. I found it unusual that these two derivatives were very easily hydrolyzed and that the methyl ether was more soluble in water than the free hydroxyl compound. To my further surprise, hinokitiol remained unchanged even on heating at 250–300 °C with concentrated sulfuric acid or 50% aqueous potassium hydroxide. Because these highly stable and amphoteric properties, as well as other experimental results, could not be explained in terms of the structural formula **27** or **32**, the real structure of hinokitiol remained a puzzle to me for some time, despite its very simple composition.

Fortunately, Pauling's book[37] became available around this time. This book was probably one of the last academic publications imported into Formosa during the war. While I was reading this book, a thought flashed in my mind that hinokitiol could be a new type of aromatic compound stabilized by resonance, with the two structures, **33a** and **33b** (somewhat similar to Kekulé's benzene formula) contributing to the resonance hybrid, and having an intramolecular hydrogen bond. Thus, I came up with the structural formula generally represented by **33c**. I presented this concept at the local Formosan branch of the Society of the Chemical Industry in 1941, but many people remained skeptical about this structure.

After 1937, European chemical journals became unavailable in Formosa, and by 1941, even the American chemical journals could not be obtained; hence, I was left uninformed of research activities going on abroad. The war had intervened in all areas of life on a large scale. Consequently, our research, including the studies of polyterpenoids, wool wax, and hinokitiol, was disrupted.

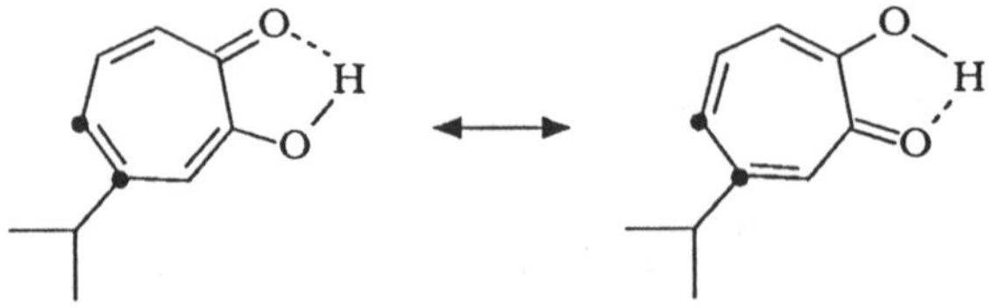

33a **33b**

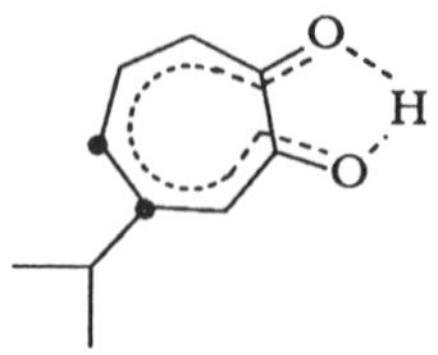

33c

World War II

The War Years (1939–1945). In February 1939, when the Japanese Army captured Hainan Island in Southern China, Formosan significance as a military base was augmented. This development further restricted our research activities. Japan's plunge into the world war with the attack on Pearl Harbor on December 7, 1941, forced us to abandon any fundamental research. Our students were drafted to assist me in research for the government (especially on natural mineral oils produced in Southeast Asia), because our previous work on natural products was not considered to be an appropriate contribution to the war effort.

From the autumn of 1942, the United States increased its bombing of Formosa. On May 31, 1943, when I was visiting the Central Research Institute, the air raid alarm sounded. I took shelter under the floor of the laboratory. Bombs came crashing down one after another—such deafening sounds as I had never heard before. My hat was even blown off in the blast!

After I emerged from the floor, I was told that several people had been killed in an air raid shelter at a girls' junior high school next to the institute. Nearby, I saw a large building of the Formosan Government-General on fire. I was told that 1-ton bombs, targeted on the central tower of the large five-story building, were dropped one after another. The bombs exploded at the base of the tower, and some part of the building was buried in rubble. The bombing is said to have killed several hundred people. The same area was hit again later that day. In the evening, when I returned to the university, I discovered that many of the buildings there had also been partly damaged and that several people had been killed by shell fragments. We realized we had to evacuate.

We could not bear to give up our research completely, so we decided to move our organic chemistry laboratory to an orange grove at Seifuku (Ch'eng-Fu), about 20 km from the city. We disassembled the laboratory and prepared to transport the instruments, glassware, chemicals, test samples, etc., to the evacuation site by wrapping them in cajuput tree bark because of the shortage of paper. The equipment was transported in army trucks for the first 18 km, but our journey was often brought to a halt by air raid sirens. For the remaining 3 km, students and laborers worked together to push flatcars up the hill. After several months, a temporary building, serving as both laboratory and residence, was completed. The day before we were to move in, the war came to an abrupt end, and an agreement was made to return Formosa to China. At noon of the same day, on an emergency order of the government, we listened to the radio broadcast. After the broadcast called the "Emperor's voice," the General Staff Office explained that Formosa was to be returned to China on the same day. Since then, the country's name has been changed from Formosa to Taiwan (Republic of China). I was quite shocked to hear this news. We had been so involved in rebuilding our laboratory that we were completely unaware of the political situation. Fortunately, no troubles arose immediately after the end of the war.

With the end of World War II, the Taihoku Imperial University was renamed the National Taiwan University under the Republic of China. Several professors of both the Faculties of Science and Agriculture were asked to help reorganize the university. Before the Taihoku Imperial University was closed, we granted a doctorate degree to Kinugasa for his structural studies on oleanolic acid and hederagenin and to Liu for his research on specific fatty acids of wool wax. One year later, Dr. Kinugasa returned to Japan, but repatriates were not allowed to take research data or information out of Taiwan. His thesis and experimental data were left to the Chinese authorities at the university, and there is no longer any record of them. To my regret, therefore, I do not remember the details of the experiment except those in the communication published in *Nippon Kagaku Kaishi* and those in the abstracts of lectures given at the annual meeting the following year, which were also published in the same journal.

Fortunately, the administrative personnel (President, Tsung-Luo Lo) who took over the Taihoku Imperial University were all very civil to those of us who remained in Taiwan at the request of the Chinese government. They let us stay at the private houses or official residences in which we had lived before the war, allowed the use of the offices at the university, and guaranteed our livelihood. All had come from mainland China, were fluent in Japanese, and had Japanese doctorate degrees.

The new Taiwan University initially proposed that they would like to discharge all the Japanese graduates, along with the Formosan technicians, and even assistants. This proposal included the chemistry department of the Institute of Tropical Medicine (of which I had been director and which was thereafter named the Research Institute of Tuberculosis) and university affiliations. I could not accept their proposal for fear that the graduates would not only lose their alma mater but also have great difficulty obtaining jobs in Japan if they no longer had supervisors. I also firmly opposed the discharge of the Formosan staff, technicians, and helpers who had been with us faithfully for so long. Fortunately, the university agreed to my request, and research continued with the same staff. The authorities told me that this special arrangement had been made "for the sake of the scientific development of Taiwan," but I think they also hoped that *l*-rhodinic acid and hinokitiol would soon be available for use as medication. Thus, I agreed to cooperate within the conditions of the new university. Remaining with me were the young graduates and Dr. Eigai Sebe, who had been appointed a professor of Taihoku Imperial University just before the end of the war.

The reconstruction of the laboratory we had abandoned during the war was not an easy task. In wartime, we were able to use the army trucks for evacuation, but naturally, these resources were not available after the war. The university did not call on us to account for our dismantling of the laboratory and evacuation of its facilities but made known their intention to take over all the properties of the former Taihoku Imperial University within the premises.

In reestablishing the evacuated facilities, we had to look after every detail by ourselves, including manpower, transportation, and costs. To secure more than 10 trucks for transportation was the most difficult part. This task was next to impossible, even for the administration of the Taiwan University. However, Professor Sebe had many influential Taiwanese friends and students who had once been with him at Taihoku Higher School. Thanks to these people, we were able to secure the necessary number of trucks. We covered some expenses by selling various items, for example, the furniture from our residences. The manpower needed for dismantling the temporary laboratory, packing, and transportation was supplied by many young students and technicians, both Japanese and Chinese, who volunteered with selfless devotion. In this respect, everything flowed smoothly. We were able to bring back almost all of the facilities to Taipei and hand them over to the Chinese administrators (Taiwan University), although some parts were broken or lost during the move, which was, after all, inevitable. Naturally, all the samples in flasks or small tubes, which we had been examining, were lost, as we terminated the research under such difficult and trying circumstances.

The new National Taiwan University kindly bore all the costs to repair and remodel the broken laboratory. Thus, after a year of hard work, we were finally ready to resume our research in a reconstructed laboratory.

Research Problems at the National Taiwan University. We initially had great difficulty choosing a research theme. In the moves to and from the evacuation site, some of our instruments had been broken, and research materials and solvents were very difficult to obtain. Under these circumstances, it was impossible to study triterpenoid structures or wool fat constituents. In the midst of these problems, I suddenly remembered the reddish brown mud that had been dumped in the yard of the Takasago Perfumery shortly before the end of the war.

One day in 1944, I had been walking down a street in Taipei and had smelled the peculiar odor of *hinoki* oil. I asked a soldier in a car nearby about the smell and was told that they were using *hinoki* oil as a gasoline substitute because of the wartime shortage. The soldier told me that *hinoki* oil, although efficient as a fuel, corroded the engines of the cars. I advised him to wash the oil with alkali. I later visited the Takasago Perfumery and was surprised to find drums of the *taiwanhinoki* oil in the yard. Years earlier, when we were studying the components of this oil, it was unavailable commercially because it was thought to be useless, and we were forced to go up to the mountains to get the material by ourselves and to extract the oil by steam distillation. At that time, even 10 L of *hinoki* oil was very precious to us. Toward the end of the war, the Japanese army had ordered the Takasago Perfumery Company to manufacture the oil from old stumps of the *hinoki* trees that remained in the mountains. They had intended to send the oil to Japan for use as a flotation oil in mines, but almost all the transport ships were struck by American torpedoes so that none of the cargo could be sent to Japan. In the backyard of the factory, I found that the reddish brown mud had been dumped on the ground.

I assumed that the unintended dumping must have happened in the following manner. Because *taiwanhinoki* oil contains 0.1–0.2% of hinokitiol, the drums corroded, and hinokitin was produced. Then, when the *hinoki* oil was washed with alkali, hinokitin precipitated with ferric hydroxide, because of its low solubility in neutral solvents such as those of pinene (main constituents of the oil). I was ecstatic to see such a large quantity of precious research material discarded in front of me so long after I had had to give up the study of hinokitiol!

Professor Sebe's Study on Hinokitiol during the War. When we began our study on hinokitiol in 1946, our colleague, Professor Eigai Sebe, told us an interesting story. In 1943, when he was a professor of chemistry at Taihoku Higher School, Professor Sebe had reexamined our study. Sebe believed that compounds containing an unsaturated seven-membered ring would not be stable enough to exist as natural products. Sebe had a different situation; he had no students, because they were all serving as student soldiers. Thus, because he gave no lectures, he had enough time to visit the Takasago Perfumery and obtain a great deal of hinokitiol, which was being prepared for Professor Katsura's chemotherapeutical study. Hinokitiol was recognized as an effective compound against tuberculosis, as a highly useful drug in the treatment of athlete's foot, and for surgery, such as operations on soldiers suffering from tetanus. In those days, no effective antibacterial agents (e.g., penicillin) were in use in Japan. Therefore, a large-scale preparation of *l*-rhodinic acid and hinokitiol was also urgently requested from the Takasago Perfumery Company by the Formosan eastern commander, General Tadao Kobayashi. He had been encouraging our studies on *taiwanhinoki* constituents while an army faculty member of Taihoku Imperial University.

Because he doubted the structure I proposed for hinokitiol, Professor Sebe reexamined this compound. He confirmed that the diol obtained by the catalytic reduction was, as we had considered, an isopropyl-1,2-dihydroxycycloheptane (**35**). However, on Clemmensen reduction of hinokitiol followed by oxidation of the resulting unsaturated hydrocarbons (**36a** and **36b**) with $KMnO_4$, he obtained a mixture of phthalic anhydride (**37a**) and isophthalic acid (**37b**). From these observations, Sebe assigned the structure **34** for hinokitiol. The natural existence of caran-type compounds was already known. Basically, he thought that the cycloheptane ring of **35** was derived by the opening of

OH OH (a) H_2 / Pt b c a =O OH (b, c) Zn/HCl

35 **34**

CH_3 $KMnO_4$ COOH HOOC

36a,b **37a,b**

the unstable cyclopropane ring at a (**34**), and the opening at b and c would result in the formation of the cyclohexadiene mixture, **36a** and **36b**. Moreover, Sebe also obtained various halogen and nitro derivatives. He thought that these compounds were the substitution products with benzene rings formed by the rearrangement of norcarane skeletons. Yet, when Professor Sebe showed me the yellow- and orange-colored crystals, I immediately thought that they were substitution products of the seven-membered ring.

National Taiwan University (1946–1948)

Hinokitiol, a New Aromatic Compound. Although Professor Sebe by no means agreed with my seven-membered structure, he willingly agreed to cooperate in our study of hinokitiol. Hoping that the mud was still there, I took a few of my research workers to the Takasago Perfumery Company. We found the factory guarded by Chinese soldiers, who would not allow us to remove the red mud. They suspected that something important was hidden in the mud—perhaps some war-related secret objects! We were dejected and returned to the university. We repeatedly asked the officials there to negotiate with the Chinese army. Finally, the negotiations were successful, and we were given all of the red mud. Everyone worked very hard and succeeded in obtaining several kilograms of hinokitiol crystals! This was a huge amount, of which prior to this time we had never dreamed. We considered ourselves exceedingly fortunate to obtain such precious research material at a time of such shortages.

Just before the war had ended (around 1943), the authorities asked Professor Shigehiro Katsura to take up a professorship at the Medical Faculty of Tohoku Imperial University in Sendai. However, to continue research on *l*-rhodinic acid and hinokitiol, he declined this offer and remained in Formosa. After the war (in 1947), he was invited to Kumamoto University as a professor, an offer that he accepted. As he was to leave for Japan, I presented him with 2 kg of hinokitiol for his medical research, but about half of this material was mistakenly stolen after he returned to Japan. It was generally known that the repatriates from Taiwan always brought back large quantities of sugar (instead of money), which was very scarce in the postwar days in Japan, and someone may have mistaken the hinokitiol for sugar. Professor Katsura even advertized through the press in hopes of the research material being returned, but this attempt was in vain. The disappointment he must have felt was beyond my imagination.

At about the time we were to resume our studies on hinokitiol, we received the first shipment of mail from Japan since the end of the war. After the war, mail was delivered only once or twice a year. I was surprised to find in the mail reprints of our study on hinokitiol,[38] which had been presented in Tokyo in 1940.[35] A coauthor of our paper, Dr. Shigeo Katsura, had submitted the paper for publication, although he left the previous (wrong) structural formula 32 untouched, because he was in favor of this earlier formula. Considering the circumstances in those days, it was inevitable that the paper would be published before he could consult with me.

Moreover, I was astonished to receive in the mail the medal for the Majima Award for Organic Chemistry from the Chemical Society of Japan without any diploma or a letter. (After I came back to Japan, I learned that this award had been given to me at the annual meeting of the Chemical Society of Japan in April 1944 for my research on natural products, sapogenins, wool wax, and hinokitiol, although full papers could not be published in wartime.) When I found the medal in the mail, I was just starting my research on hinokitiol, so the prize was all the more encouraging for me and my co-workers. I have heard that this particular shipment of mail had been on its way to Formosa when the transport ship had been attacked and had sunk just outside of Keelung (Telong) harbor; after the war, the ship was salvaged, and the mail was delivered to us.

In the summer of 1946, we finally began our research on hinokitiol. A large research group was formed, consisting of Professor Sebe; three Chinese associate professors, P.-Y. Yeh, F.-C. Chen, and L.-S. Chen; four Japanese instructors, Y. Kitahara, S. Mayama, S. Iwamoto, and H. Fujii; and several Chinese students, T. J. Hsü, L. C. Tseng, L.-C. Lin, and T. B. Lo. Dr. K. Pan, an associate professor of physical chemistry, also assisted us in a study of the physical properties of hinokitiol and its complex salts. Later, Y. T. Lin (a former student) joined us as an associate professor after he returned from mainland China. Dr. S. L. Liu, whose studies in the United States I helped to arrange, came back to Taiwan after my repatriation. All these people were later appointed as full professors of the National Taiwan University.

We began our study with the careful reconfirmation of the functional groups of hinokitiol. The formation of the neutral monoacetate and monomethyl ether indicated that one oxygen atom belonged to the acidic enol group. The saturated ketone ($C_{10}H_{18}O$) was found among the catalytic reduction products, although in very minute amounts. With di- and trivalent metallic ions, hinokitiol produced neutral complex salts having the structures **38a** and **38b**. A complex salt of the type **38c** was also obtained by Dr. Hiroshi Iinuma (*see* The Situation in Japan,

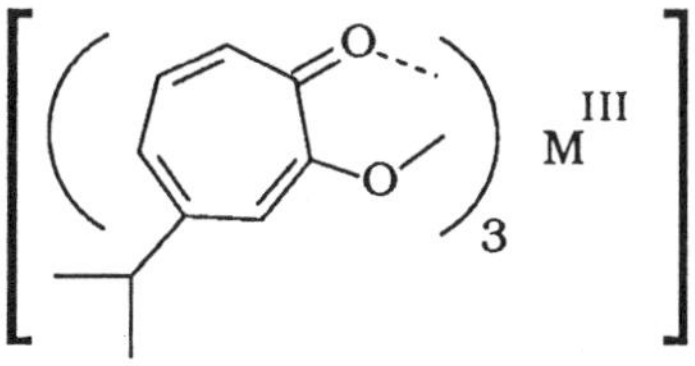

38a

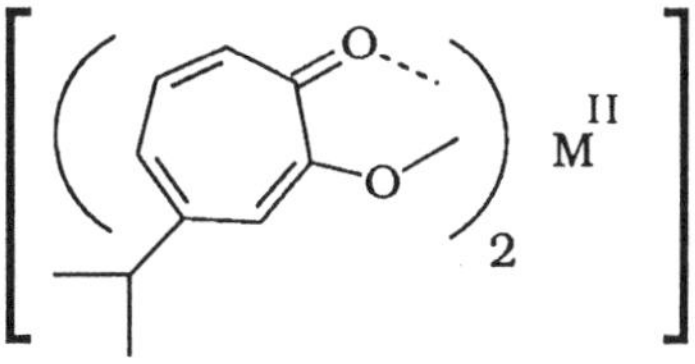

38b

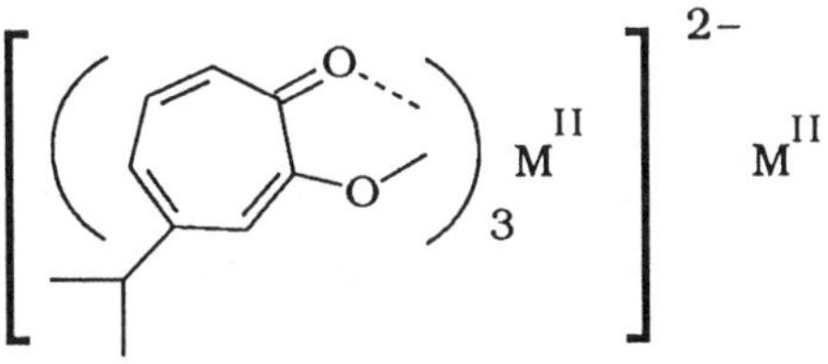

38c

to follow). The formation of complex salts suggested that the other oxygen atom belonged to an inactivated carbonyl group.

We encountered extreme difficulty in determining the number of double bonds in hinokitiol. On addition of bromine to a solution of hinokitiol in ethanol or acetic acid, we observed the liberation of HBr, suggesting that the reaction taking place was not a simple addition reaction. During catalytic reduction over PtO_2, 4 mol of hydrogen was absorbed, and a 60% yield of the diol (35) was produced, along with a monohydroxylic compound ($C_{10}H_{20}O$), a ketone ($C_{10}H_{18}O$), and a hydrocarbon. The molecular refraction (Lorenz and Lorentz formula) of hinokitiol was also examined. We were greatly excited to find that hinokitiol has an unprecedented and unusual system of conjugation. The substance remained unchanged by catalytic reduction over Pd–C or by Bouveault–Blanc reduction, a fact indicating that the double bonds were relatively stable.

Next we examined the substitution reactions of hinokitiol. It is no exaggeration to say that these investigations brought continuous surprise to us. One mole of bromine in an ethanolic solution of hinokitiol at low temperature produced a neutral oil that was considered to be a bromoethoxy derivative, in addition to 1 mol of HBr. On treatment with an alkali, this oil quantitatively produced the sparingly soluble sodium salt of monobromohinokitiol. In acetic acid, more than 2 mol of bromine was absorbed, and the products were isomeric mixtures of mono- and dibromo derivatives, as well as a small amount of the tribromo compound. All of these substances formed sparingly soluble sodium salts and metal complexes with ferric chloride. We thus confirmed that these substances were the substitution products still possessing the seven-membered ring. Because hinokitiol easily gave a trichloro compound (39, $X^1 = X^2 = X^3 = Cl$) by the reaction with chlorine

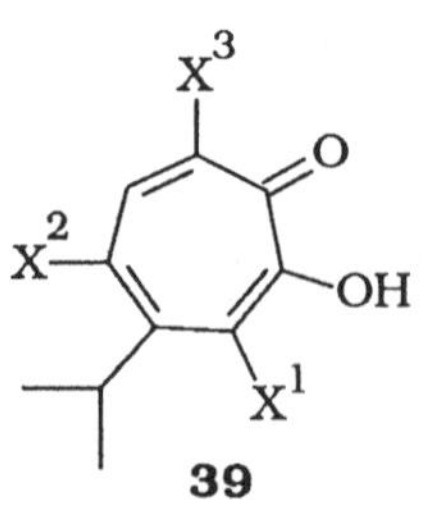

X^3 O X^2 OH X^1

39

in acetic acid, the isopropyl group was considered to be at C-4, corresponding to the *meta* position. Surprisingly, we found that the trichloro compound could absorb an additional 2 mol of chlorine to become a pentachloro compound. It was then transformed into a carboxylic acid and a phenolic substance.

Members of the nitrohinokitiol research group at the department of chemistry, National Taiwan University in Taipei, 1947. From left: L.-C. Lin, Yoshio Kitahara (instructor), T. B. Lo, Miss L. C. Tseng, Miss Yoshiko Lin (secretary), and Tetsuo Nozoe.

Nitration turned out to be more complicated; reaction with concentrated nitric acid yielded nothing at temperatures lower than 10 °C. However, at room temperature, once started, the reaction progressed rapidly and exploded occasionally. When the temperature was lowered immediately after the initiation of the reaction and maintained at 20–30 °C, an almost quantitative yield of the mono or dinitro derivative was obtained, depending upon the amount of nitric acid used. The dinitro compound was believed to be the 5,7-dinitro derivative **40**. With arylamine in benzene, the dinitro compound gave a 1:1 salt. On addition of one drop of ethanol, a hydroxyl group in the salt was replaced by an arylamino group to give **41**.

41 ← **40** → **42**

When we heated the dinitrohinokitiol **40** at 50 °C in ethanol for recrystallization, it suddenly boiled and became a colorless, neutral substance with a low melting point. Similar results were obtained with common primary and secondary alcohols. Heating with aqueous acetic acid produced the carboxylic acid **42** (R = H), and we confirmed that this neutral substance was an ester (**42**, R = CH_3, C_2H_5, etc.). We reasoned that the product was formed by a benzylic-type molecular rearrangement, but we had never seen an example of this rearrangement under such mild conditions. With *o*-phenylenediamine, the reaction produced a quinoxaline derivative. Yet, when we tried to obtain a phenylhydrazone by treating **40** with phenylhydrazine, the reaction mixture suddenly exploded. However, compound **40** was a stable substance at room temperature.

Hinokitiol gave a quantitative yield of orange-colored arylazo compounds, **43**, by the usual method. When **43** was heated in ethanol or acetic acid, it gradually changed into dark-violet quininoid compounds, **44** (which we called hinopurpurin), with a metallic luster and gave quinoxaline derivatives, **45**, with *o*-phenylenediamine. The exact structures of these compounds (**39–45**) were established upon our return to Japan, as will be described later (*see* S and N Analogues of Tropones, Heptafulvenes, and Quinarenes, to follow).

The experiments just described proved my hypothesis that hinokitiol was an entirely new type of aromatic compound. However, I was surprised to learn that this compound was quite different from the ben-

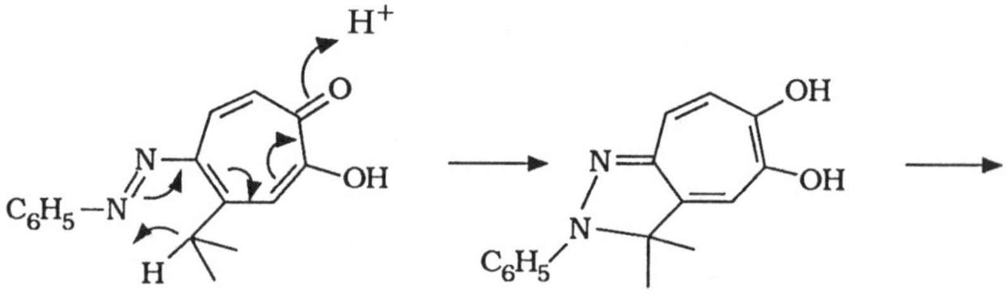

H^+ O OH C_6H_5—N N H → N N C_6H_5 OH OH →

43

O O N N C_6H_5 → N N C_6H_5 N N

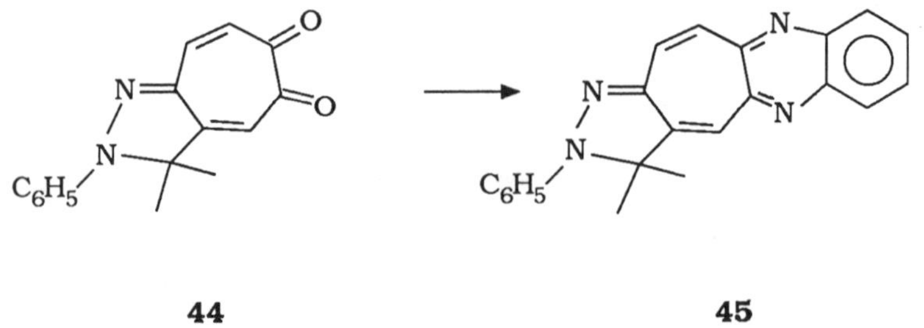

44 **45**

zene series. In the following year, after I returned to Japan, a review on our study in Formosa was published.[39] However, formula **33** for hinokitiol had been previously cited in medical journals by my co-worker, Professor Shigehiro Katsura,[40] and most of our experimental details on hinokitiol at National Taiwan University were published in 1951–1952.[41]

Repatriation to Japan. Our studies at the National Taiwan University were surprisingly productive, considering the conditions directly following the war. I was not to know, however, the postwar events in Taiwan. One by one, all of the Japanese people working as instructors at the National Taiwan University went home; even my young son had returned to Japan alone. Finally, only Professor Sebe; my assistant, Mr. Kitahara; I; and our families remained. (Yoshio Kitahara was born in Amoi in mainland China, where his father was a teacher at the Japanese school. He was my closest colleague during the early period of my troponoid studies at Formosa and Tohoku Universities.)

I felt responsible for the graduates of the Taihoku Imperial University, who had lost their alma mater and were left unemployed. I also felt that if we retained our posts at Taiwan University, we might be preventing the Chinese associate professors from assuming full professorships. For these reasons, I felt that we should return to Japan. We submitted our resignations to both the president and the dean of the university, who refused to accept our resignations but promised to cooperate with us and assist us with our research activities if we stayed in Taiwan.

Early in 1947, an unfortunate incident, the "February 28th Incident," occurred, and our laboratory was temporary closed. Some Taiwanese people (especially the soldiers who had fought in the Japanese army and some citizens) rebelled against the Chinese government. Again, we attempted to turn over our posts to Chinese chemists and go back to Japan. The authorities at Taiwan University still would not approve our resignations and cited their inability to find suitable successors and the possibility that our resignations might cause unrest among the Chinese students.

Early in 1948, the president of Tohoku Imperial University in Sendai wrote me a letter requesting my return to Japan to fulfill an appointment as a professor at Tohoku Imperial University. I also received a letter from Dr. Kafuku, my former supervisor, and I was surprised at the tone of his letter, in which he commanded me to come home: "You are neglecting your own son and failing in your duty to the graduates of the former Taihoku Imperial University, who are still without jobs. Do you intend to stay in Taiwan forever?" From his letter, I learned that my son, who had been sent to stay with my elder brother in Osaka, was now living with Dr. Kafuku in Tokyo, because my brother's house had been requisitioned by the occupying army. I showed these letters to the president of Taiwan University, Chih-Hung Lu, and again requested his approval for my return. He finally accepted our resignations; moreover, he made an unusual exception for us, by allowing us to take the hinokitiol sample and its derivatives for our future research in Japan. This gesture was a token of gratitude for our long service and our contribution to the establishment of the Chinese National Taiwan University.

Thus, we were able to bring our precious research material back to Japan with us, although we had to leave behind many of our personal belongings to make room in our luggage. Repatriates were allowed to bring back only the minimum living necessities and 1000 Japanese yen. As a general rule, valuables and houses were confiscated. Of course, research data were not considered an exception to this rule. However, President Lu kindly drew up a certificate for us so that we could return to Japan with a large quantity of hinokitiol and its derivatives. We were greatly indebted to him for this favor. To return this kindness, almost every year I invited former students or colleagues from my Formosa days to Tohoku Imperial University in Sendai at my own expense and obtained for them Rotary Club fellowships for the next year. Eight of these students were granted Ph.D.s from Tohoku Imperial University. Without the kind cooperation of the Taiwan University authorities and the assistance of the co-workers and students, our later study of troponoid chemistry in Japan would not have been possible. For my work in Taiwan and my interest in helping Taiwanese chemistry, I was awarded the Medal of Culture from the Ministry of Education, Republic of China in 1978 and presented with an honorary citizenship of Taipei in 1982.

Along with our families, we left Taihoku in early May 1948. Many of the faculty and students of the university kindly saw us off. We were placed in a reception center for repatriates in Keelung while we awaited the ship that would take us to Japan. The center was quite a contrast to our official residences and laboratories at the university; it was dirty and uncomfortable, and during our 2 weeks there, all we did was practice packing and unpacking our luggage in preparation for inspections. Finally, a former Japanese training ship, *Kaio-Maru*, arrived. Many staff members (who have since become professors at Taiwan University), students, technicians and helpers came to say good-bye at Keelung. They started to sing a farewell song in Japanese, but were stopped by military policemen with bayonets, because the Taiwanese were forbidden from speaking Japanese in public. I was filled with mixed feelings. I was sad to leave Taiwan, which had become my

The organic chemistry group, including members of the Research Institute of Tuberculosis, gathered in front of the National Taiwan University chemistry building in Taipei, March 1948, for a farewell party for the Japanese staff, who were being repatriated to Japan. First row from left: F.-C. Chen (associate professor), P.-Y. Yeh (associate professor), D.-C. Chung (professor, chairman of the department of chemistry), Tetsuo Nozoe (professor), E. Sebe (professor), T. Kinugasa (associate professor), Y. Kitahara (instructor), L.-S. Chen (associate professor), and I. Tominaga (Research Institute of Tuberculosis). Second row, third from right: Y. Endo (head of the glass shop) behind Professor Kitahara and wearing glasses, and Miss Yoshiko Lin (secretary to Nozoe), far right. Behind her in the third row are Miss T.-L. Lin (librarian), and Miss S.-M. Lin (secretary to Sebe).

second home, after I had spent 22 years of my life there. It was especially painful to part from the Taiwanese staff, students, and technicians who had been so very kind and helpful to me. On the other hand, mixed with this sorrow was a strong desire to return to Japan and become active in organic chemistry, now that we had the research on hinokitiol as a foothold and, more specially, to reunite with my son, relatives, Dr. Kafuku, and many friends.

An alumni meeting of National Taiwan University (Taipei, November 1987) 40 years after repatriation to Japan. Most of the staff members, technicians, and helpers from those days attended the reunion, some together with their families.

The ship stopped at Shanghai and Tsingtao to take on more Japanese repatriates and then finally arrived at Sasebo, Kyushu, on May 25, 1948. I became ill from the uncomfortable conditions at the reception center in Taiwan and the long sea journey and had to spend a few days in the hospital of the reception center in Sasebo. We boarded a special train for repatriates and traveled the 2-day journey to Tokyo. My family and I then joined my son and settled with my sister, whose house in the suburbs of Tokyo had survived the bombings.

The day after our return, Dr. Kafuku came to see me in my sickbed. After asking about my research in Taiwan, he encouraged me to try to contribute to the scientific world through the study of the peculiar chemistry of hinokitiol. Dr. Kafuku apologized for the sharpness of the letter he had written, in which he demanded my return to Japan, and explained that it had been strategically written after he consulted with the faculty of Tohoku Imperial University to facilitate my repatriation.

I was shocked to receive a telegram 10 hours later saying that Dr. Kafuku had died. Unfortunately, I could not attend his funeral because of my sickness. I learned later that a fire had broken out that night in the laboratory where he worked and lived with his family and that he had had a heart attack while trying to put out the fire. I made a vow to promote a study on hinokitiol and its allied compounds in payment for the kindness of Dr. Kafuku and others.

Tohoku University (1948–1966)

Preparations for Research

The Situation in Japan. After the war, repatriates first had to be inspected severely before being appointed to serve as a government official. This process usually took several months. In July 1948, I was appointed a temporary lecturer of the Department of Chemistry of Tohoku University until inspection was finished.

Eventually I recovered my health, and in October 1948, I began preparing for our study at the Department of Chemistry at Tohoku University in Sendai. Besides Yoshio Kitahara, who had returned from Taiwan with me, S. Seto, K. Kikuchi, T. Mukai (students of my predecessor, Professor Hiroshi Nomura), T. Ikemi (a graduate of Hokkaido University), Kameji Yamane, and Shô Itô (my first students at Tohoku) also joined me.

In 1948, Shuichi Seto, an officer at the Navy Fuel Bureau during the war, joined the group as an associate professor at Tohoku University. With my assistant Kitahara, Seto became immersed in the resurrection of our laboratory during the most trying period of postwar Japan. Seto was responsible for taking care of students and constructing a small "factory" to synthesize tropolones.

By this time, Japan was gradually recovering from the disorder of the war, and the food shortage had somewhat improved. Sendai City had been heavily bombed during the war, and the main sections of the city were burned. Upon my arrival, I was very sad to find that the house my family had lived in when I was young had been destroyed during the war. Most of Tohoku University had survived the fires, but it still had suffered a great deal of damage. We found school personnel and students who had lost their residences during the war still living in the chemistry building, and research activities had not yet resumed. The situation in Sendai seemed to be worse than that in Taiwan.

The results of our study on hinokitiol conducted in Taiwan had been disclosed to some chemists in Japan by students who had preceded us. Some professors at the Department of Pharmacy at Tokyo University who had heard of our study thought that the norcarane structure **34** would be more appropriate than the trienolone **33** for hinokitiol. In August 1948, Professor Sugasawa of Tokyo University received a letter from a colleague in Sweden, Professor Erdtman, asking for the names of the chemists who were studying the natural compounds called hinokitiol, which had a similar seven-membered ring structure to the thujaplicins[42] he was studying. In Taiwan, I had asked my colleague, Dr. Iinuma (an assistant professor of inorganic chemistry), to study the metallic complexes of hinokitiol. Professor Erdtman had seen Iinuma's paper on the metal complexes of hinokitiol, **38a–38c** in *Chemical Abstracts*.[43,44] I was surprised to find that such a similar study[42] had been conducted overseas and had been published around the time I returned home. However, Dewar's reports,[45,46] which were cited in that paper, had been published in 1945 in *Nature* but were unavailable in Japan at that time.

Toward the end of that year, I was invited by Professor Majima (who by then had retired as president at Osaka University) and Professor Kotake of the Department of Chemistry to a special seminar of the Chemistry Department at Osaka University. I gave a 3-hour lecture on our study of hinokitiol in Taiwan and showed various samples of hinokitiol derivatives and colorful hinopurpurins. The following February, I was also invited to give the same lecture at Tokyo University.

The chemistry department at Tohoku University in Sendai at the time of Tetsuo Nozoe's retirement in 1966. The right side of this building, which housed the laboratories, was built in 1935; it escaped damage from fires during World War II. The occupation forces wanted to use it as a hotel, but the chairman of the chemistry department opposed the plan vehemently. The left side of the building was added later. (Photo courtesy of M. Yasunami.)

After a detailed explanation, the structural hypotheses (**33a–33c**) were accepted, and I received the approval and encouragement of both faculties of Osaka and Tokyo Universities. At the request of the professor in charge from Tokyo University, a draft of my lecture was printed without abridgement in a special temporary review journal.[39a] Because the newly organized Chemical Society of Japan had just started to publish only short papers (each two pages) every month, I was extremely fortunate to have my extensive paper on hinokitiol published in this special review journal.

In December 1948, the same day I was appointed a professor of Tohoku University, Professor Roger Adams (University of Illinois), leading an investigation into the chemical world of Japan after the war, visited Tohoku University and gave a lecture on lupine alkaloids. It was the first time since the war that we had been informed of a purely scientific and new research result. I spoke to Professor Adams about my study on hinokitiol and was encouraged by his interest. After his lecture, we talked about chemistry long into the night; I realized that Professor Adams loved science and encouraged young chemists. I became more confident about my study, and Roger Adams and I became close friends.

Professor Adams never failed to call on us in Sendai whenever he had a chance to visit Japan. In 1953, when I first traveled abroad and visited his laboratories in Urbana, IL, the first place he took me to was a barber shop. He said that I must have been tired and needed refreshing. He waited for me, and together we went to the university, where I gave my lecture. In April 1964, during Professor Adams' third visit to Sendai, he gave a lecture for more than 1 hour on "Fifty Years of Organic Chemistry in the USA". The next morning, I went to his hotel to pick him up and found him suffering from angina. Because a medical conference was being held in Sendai on the very day, all the key professors of the university hospital were out for the conference. I sent for one of the professors so that Roger Adams could be treated as quickly as possible. Fortunately, he recovered well enough to return to the United States. I accompanied him to Tokyo by air to ensure his safe departure. He was grateful and told me that he wanted to come to Sendai again soon. Unfortunately, he passed away before making his fourth trip to Sendai. When he died, I was traveling abroad, and only after I returned to Japan did I hear the news. I was very saddened to hear of his passing.

The Early Stages of Tropolone Chemistry Overseas (1945–1948). It was only until after 1949 that we had access to American chemical journals from the American Cultural Center in Sendai. Later we were also able to obtain European journals, which usually arrived more than 6 months after publication. From these publications, we learned that studies on a number of naturally occurring seven-membered-ring aromatic compounds had been performed in the West, almost simultaneously with our hinokitiol study.

On the basis of the experimental data of Raistrick et al. (1932–1942), Dewar,[45] in 1945, proposed structure **46** for stipitatic acid. He proposed to call this novel seven-membered nucleus, **47**, "tropolone," which he, at first, thought to be stabilized by resonance, with

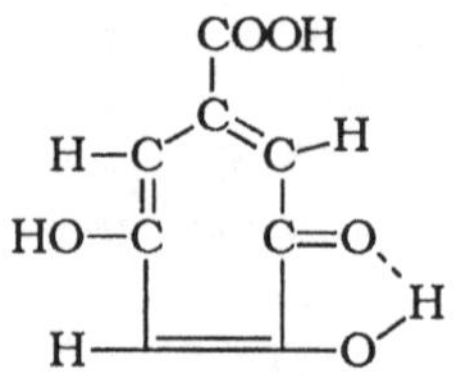

46a 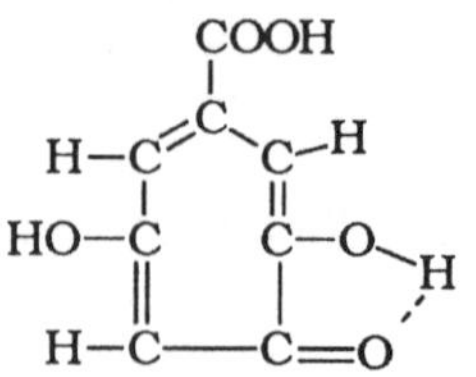 **46b**

A bit of illness and I found the usual Japanese cordiality and hospitality simply overwhelming. I have no way to express effectively my gratitude and appreciation of your attention and kindness to me. I sincerely hope that I may sometime in the future reciprocate. In spite of my confinement I enjoyed immensely my visit to Sendai.

For chemistry at my age, it should be simple and unequivocal.

C_2H_5OH. I shall look forward to my next visit

Roger Adams 4/28/64

Roger Adams, November 1964. Professor Adams had signed Nozoe's autograph book just before departing Tokyo in April of that year. The "gratitude and appreciation" were for Nozoe's support and assistance during an angina attack Adams had experienced during his stay in Japan. (For a discussion of Nozoe's autograph books, see p. 113. Photo courtesy of K. Nakanishi.)

structures **47a** and **47b** contributing to the resonance hybrid. By modifying the traditional Windaus formula, **48**, for colchicine, which was attracting immense attention because of its very interesting biological activity, Dewar[46] also proposed structure **49**, having a seven-membered

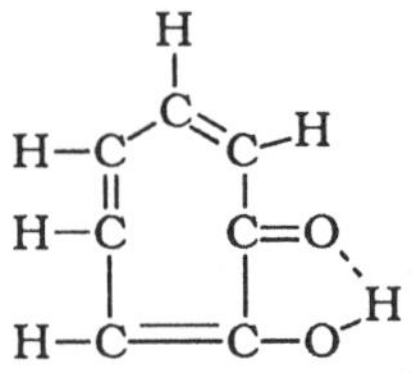

47a 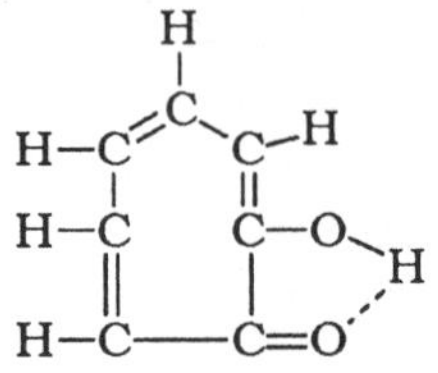 **47b**

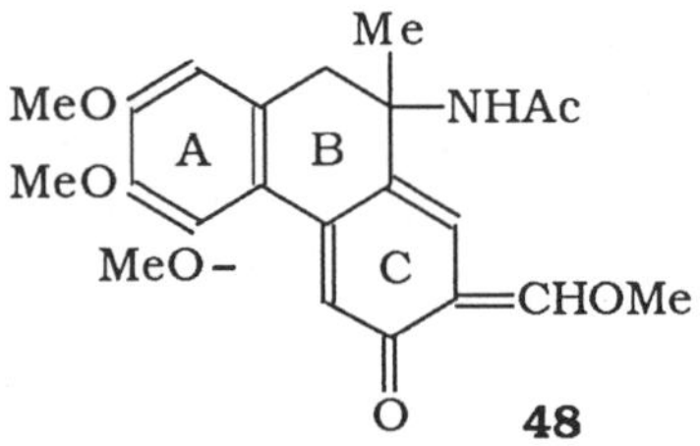

48 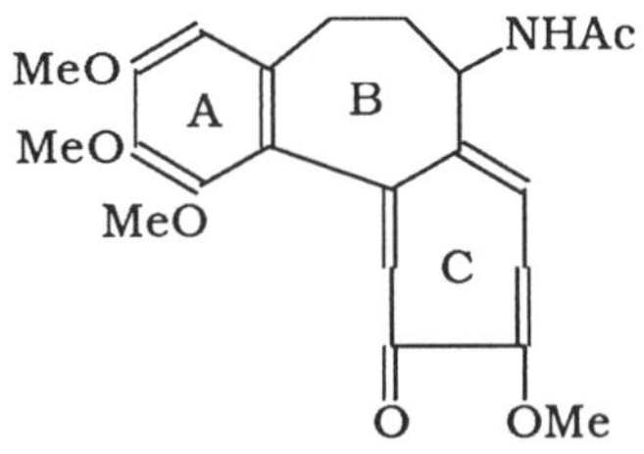 **49**

C ring, for colchicine. The structure of the troponoid C ring in colchicine was later confirmed experimentally.[47]

In 1948, Erdtman et al.[42] assigned the isopropyltropolone structures **50a**–**50c** to α-, β-, and γ-thujaplicins after a systematic degradation study. On the basis of the UV spectra and the acidities of thujaplicins, Erdtman et al. pointed out that the seven-membered ring system was stabilized through the resonance structures **51a** and **51b** with intramolecular hydrogen bonding. This was the first report on natural tropolone established as the result of experiments. Structures **46**–**51** are reproduced in the 1940s mode of depicting the concepts and structures.

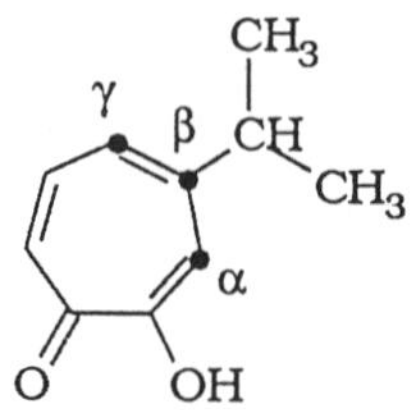

50a - c

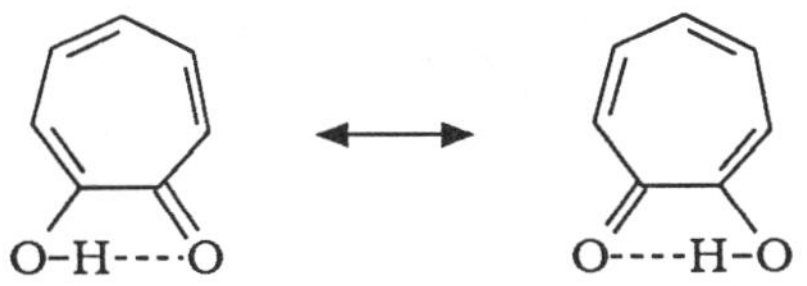

O-H---O O---H-O

51a **51b**

Very similar studies had also been conducted abroad, although almost no studies had been carried out on the chemical reactivities of these new aromatic compounds. We also noted that the benzotropolone formula (52) had been proposed independently for purpurogallin by two laboratories in England.[49,50]

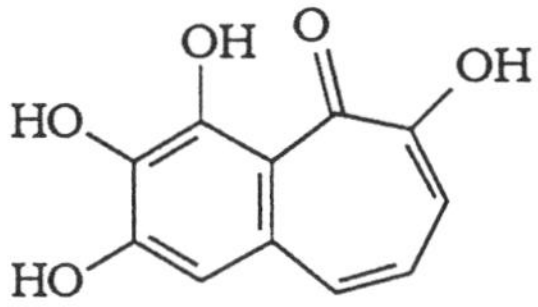

OH O OH HO HO

52

Overseas Research on Triterpenoids and Wool Wax. As the size of our research group gradually increased, I began, around 1954, reviewing papers on sapogenins and wool wax that had been published during the war. I wondered whether there was room to continue our studies in these areas, which had been discontinued for almost 15 years because of the war, and at the same time continue with our efforts on seven-membered-ring aromatic compounds.

I noticed that, in 1939, Kitazato,[51] who had been actively studying oleanolic acid and hederagenin, had adopted the structural formulae we had proposed, **12a** and **12b**,[12] instead of his own formula, **53**. He also quoted our conjugate triene structure **11**.[12,13] Subsequently, Ruzicka, a leading expert in polyterpenoid chemistry, had also adopted[52] our formula 12 instead of his own formula **54**.[10]

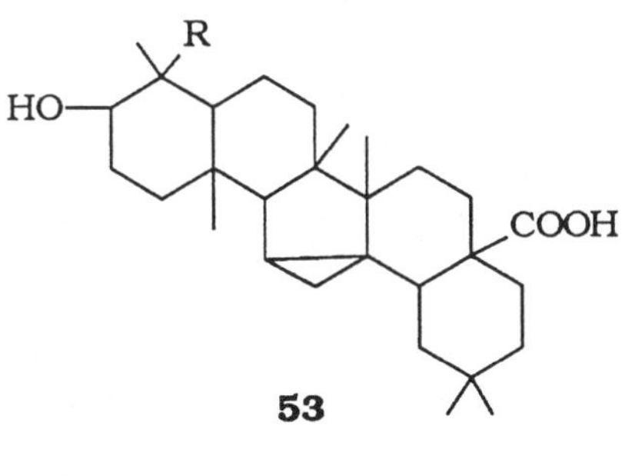

R HO COOH

53

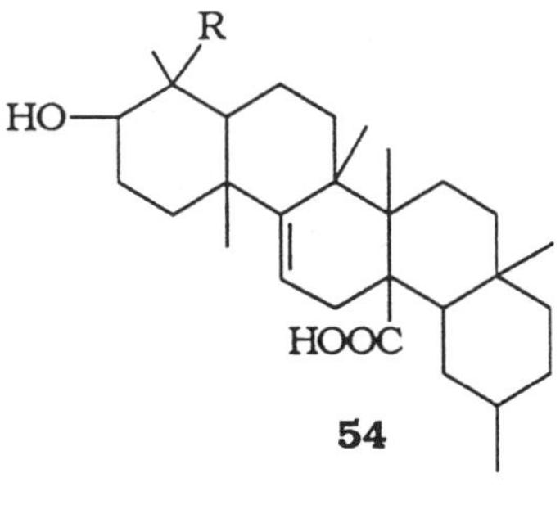

R HO HOOC

54

Dr. Michael J. S. Dewar in 1951.

To my surprise, however, many researchers[51–53] and reviewers[54,55] believed that formula **12** had been proposed by Haworth.[9] When I examined this point in Haworth's *Annual Report of Progress in Chemistry* for 1937[9] on triterpenoids, I found that Haworth concluded as follows: "This survey of the general structures of oleanolic acid and hederagenin indicates that no formula has yet been advanced which gives a satisfactory explanation of both the dehydrogenation and the oxidation results. During the remainder of this Report, formula XXIII, which explains the dehydrogenation results, will be adopted for the pentacyclic triterpenes other than lupeol, in spite of its failings in connection with the oxidation results." After this concluding statement, the reviewer added the following in the footnote: "Structure a is worthy of consideration; it is composed of isoprene units, contains the system XXIII, and supplies a rational explanation of dehydrogenation and oxidation experiments." Formula XXIII and structure a in this report[9] were identical with **54** by Ruzicka and my structure **12**, respectively. Unfortunately no author reference for this footnote was given. Because *Nippon Kagaku Kaishi* was listed among the journals reviewed

Nozoe's group at Tohoku University on graduation day, taken at the main entrance of the department of chemistry in Sendai, 1952. Front row, beginning third from left: Yoshio Kitahara, Tetsuo Nozoe, Shuichi Seto, and far right, Miss T. Sato (later Mrs. T. Asao). Second row from left: Shô Itô; 4th from left, I. Murata; T. Mukai; 7th, S. Masamune; and T. Muroi, far right. Third row, 3rd from left: S. Morosawa, M. Sato, T. Ikemi, H. Takeda, T. Arai, and K. Doi.

for this annual report for 1937,[9] it became my view that the reference[12] for this footnote must have been erroneously dropped during the printing.

Later, in a paper by Tsuji[56] on his novel decarbonylation reaction of acyl halides by Pd metal, reference was made to our structures **14** and **15**. Although the planar structure **12** was still being used without reference to our communication,[12] I felt that the chemistry of these polycyclic natural products had already entered the era of stereochemistry.[57] In 1953, I learned that in the United States and Europe, recording UV and IR spectrometers were being used frequently. Moreover, it was very difficult in Japan of those days to obtain sapogenins and solvents in the large quantities needed for research. In foreign countries, the leading organic chemists were already studying this topic energetically and had advanced 15 years ahead of us. I found that experts in natural products chemistry, such as Ruzicka, Barton, and Spring were concentrating on these topics. It was more than likely that they would solve the problem sooner or later. Thus, considering it best for me to concentrate on the tropolone research, I abandoned the sapogenin research. It was unfortunate that the western chemists (Ruzicka, Spring, and Barton) published the same formula as the one I had found without being aware of my work, because of a misprint in the annual reports.[9]

After checking the overseas research activities regarding the constituents of wool wax, I was again surprised to find that our prewar discoveries mentioned earlier had been published with the same results. Namely, in 1945, Weitkamp[58] isolated wool wax fatty acids by high-precision fractional distillation of the ester mixture and found two series of branched-chain fatty acids, **17** (iso, C_{10}–C_{28}) and **18** (anteiso, C_9–C_{27} and C_{31}), in addition to the normal acids (C_{10}–C_{26}). This finding was almost identical to ours.[23] In determining these structures, Weitkamp did not use any chemical methods, but rather, he used only the numbers of transitions appearing in the solidification point curves of the binary mixture between the branched acids (or their amides) and normal fatty acids (or their amides). In subsequent papers and monographs,[59,60] only Weitkamp's paper was quoted.

In addition to these two series of unusual fatty acids, lanoceric acid and lanocerol were also reported[59,60] without reference to our research.[24–26] However, our findings had been filed into 10 Japanese patents in 1941. Of these, only two items[26] on α-hydroxy acids were referred to in *Chemical Abstracts* and quoted in the monographs.[59,60] Furthermore, I knew that lanochromin (**22a**) and lanochromic acid (**22b**), which we had obtained by the oxidation of lanosterol, were later reported independently.[61–63] Thus, the determination of the structure of lanosterol was within reach.

Our findings from the studies on sapogenins and wool wax constituents prior to and during the war were published only in short communications or as abstracts at domestic meetings. Thus, it was understandable that our results were unknown abroad. Because of this situation, I decided to give up our research on these natural products, even though it had taken up most of my efforts over the past 10 years. However, the fact that our findings were correct, despite the lack of any modern analytical tools, did comfort me greatly. Also, because of this decision, we could concentrate all our efforts on troponoid chemistry.

The Dawn of Troponoid Chemistry (1948–1950)

Structure and Aromatic Properties of Hinokitiol Derivatives. The first task I approached after returning to Japan was the establishment of the structure of the hinokitiol derivatives prepared in Taiwan. We found that the reactions of many hinokitiol derivatives closely resembled those of the corresponding benzene derivatives, namely, a halogen atom attached to the nucleus could be removed by catalytic hydrogenation and nitro and arylazo groups could be reduced to amino groups. These were then converted into halogen or parent compounds by diazo reactions.

From these reactions, we were able to correlate various substitution products[64] as illustrated in Scheme I. By applying a known procedure to the rearranged dinitro compound (**42**), we obtained isophthalic acid.[65] From this result we established that the isopropyl group is at C-4 (*meta* position) of the hinokitiol nucleus as had been assumed earlier (Scheme I).[65] Compared with halogenation and azo coupling, the nitration of hinokitiol was more complicated and varied, and even a slight, almost imperceptible deviation of reaction conditions gave a number of different products or a large difference in yields. As a result of the repeated nitration experiments under different conditions, we isolated five kinds of pure mononitro compounds (α, β, γ, δ, and ϵ isomers). To isolate these compounds, the noticeable differences in their acidity and the ease of crystalline-Na-salt formation were exploited, in addition to the use of ordinary recrystallization methods.

This number of compounds was more than we had expected, and initially, we were puzzled by their structural determination, because their reduction produced only one of 3- (from α), 5- (from δ and ϵ), or 7-aminohinokitiol (from β and γ isomers). We first considered the existence of rotational isomers[66] (e.g., **55a** and **55b**), but later, when a

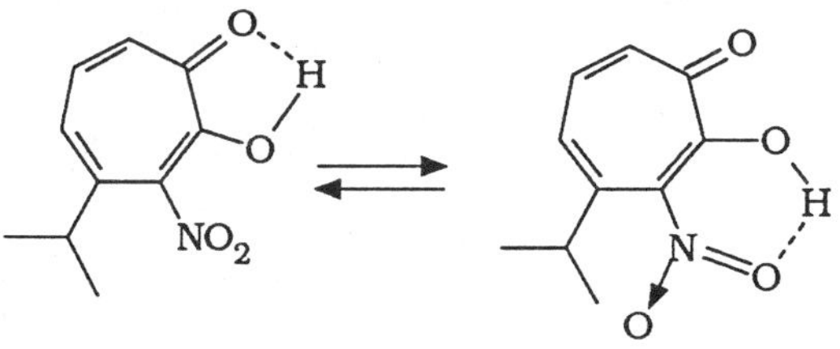

O H O NO$_2$ ⇌ O O H N O O

55a **55b**

large group was substituted at C-3 or C-5 vicinal to the isopropyl group at C-4, two prototropic isomers (α-enolone, **56a**, and diketone, **56b**) were

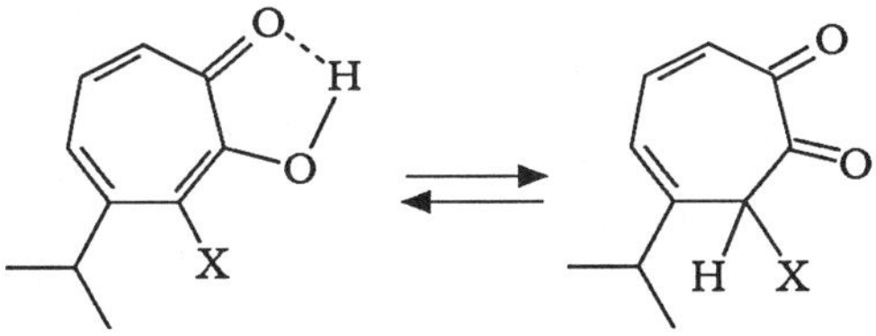

O H O X ⇌ O O H X

56a **56b**

Scheme I

usually produced because of steric hindrance of the side chain. Unlike the usual tautomers in a state of dynamic equilibrium, these isomers remained stable at room temperature, presumably because of appreciable hindrance to the free rotation of the vicinal isopropyl and nitro groups (*see* Electrophilic Reactions of Tropolones, to follow). Thus, the structures of the various substitution products of hinokitiol were determined, and the stability of the double bond system became clear.

Synthesis of Tropolones. The next important work was the synthesis of the seven-membered aromatic ring, tropolone (**47**), which was predicted by Dewar.[45] The significance of this undertaking is evident from the fact that, in 1949, when we began our synthetic research in Japan, it was difficult to obtain commercially even simple reagents; a trivial compound, such as methylamine, had to be prepared by methylation of ammonia.

Initially, we attempted the synthesis of tropolone (**51**) by dehydrogenation of 1,2-cycloheptanedione (**57**) with Pd–C, but this pro-

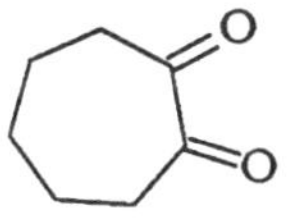

O
O

57

cedure was unsuccessful. Although the structure of tropolone looks very simple, the actual synthesis of monocyclic tropolones was difficult in the early stages. However, by the end of 1949, after treatment of **57** with NBS (*N*-bromosuccinimide) or Br_2 under appropriate conditions, we were able to synthesize tropolone and, especially, its bromo derivatives (**58** and **59**)[67] in a high yield. However, without careful con-

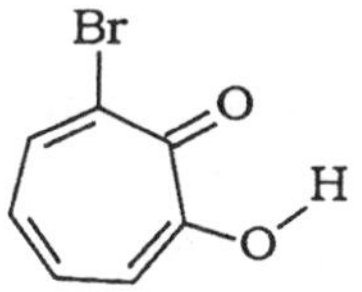

Br O H O   Br O Br OH Br

58 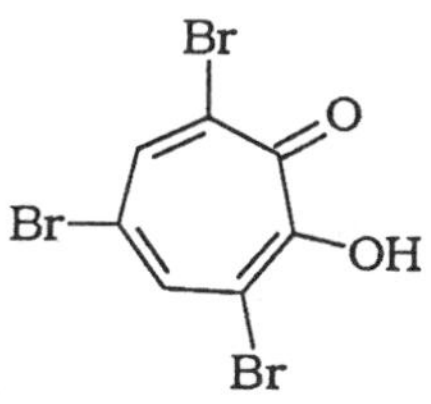 **59**

trol, this reaction often causes an explosion. Subsequently, using the same procedures, we succeeded in the synthesis of thujaplicins[68–70] as well. In contrast, the synthesis of benzotropolone (**61**) from diketone (**60**) was easily achieved by Cook et al.,[71a] using the dehydrogenation

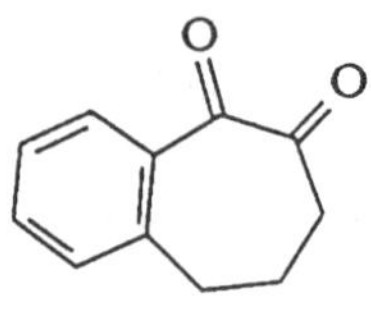

60 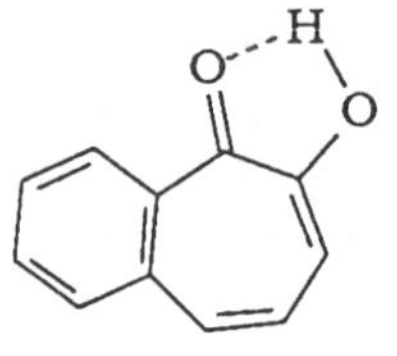 61

method just mentioned. Dibenzotropone was obtained more easily, but it exists as a diketone rather than as a tropolone.[71b]

The characteristics of the synthesized monocyclic tropolones, such as acidity, complex formation with metals, halogenation, and azo coupling, were quite similar to those of hinokitiol. On the basis of the amphoteric nature and the chemical properties of these compounds, I considered that the seven-membered nucleus was stabilized by the resonance contribution of dipolar structures, such as **62b** and **62c**, in addition to the neutral structure **62a**, which has intramolecular hydrogen bonding.[72]

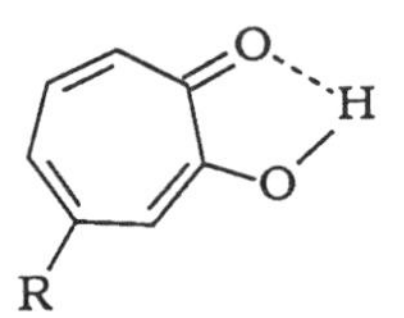

etc.

62a 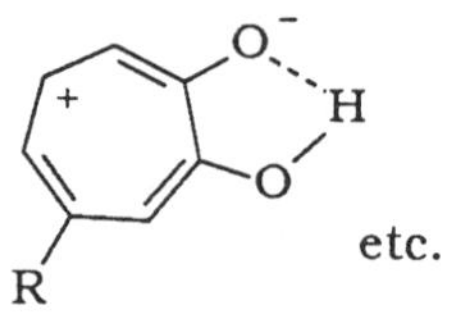 62b

etc.

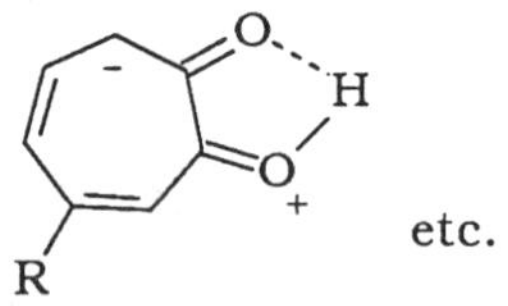

62c 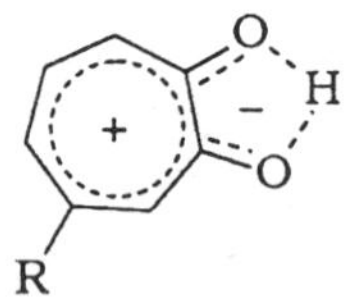 62d

Realizing that the traditional method of studying organic chemistry was not enough for the study of these unusual compounds, I asked Professor Masaji Kubo of Nagoya University for his assistance in the measurement and theoretical calculations of the dipole moment of tropolone, hinokitiol, and their derivatives. From these studies, it was obvious that the tropolone nucleus had a nearly planar heptagon with a positive charge on the seven-carbon ring, as shown by the structure **62d**.[72,73] The dipole moments enabled us to confirm the position of the substituents as well. I also requested the help of many other leading Japanese physical chemists to study the aromaticity and fine structures of these new seven-membered aromatic compounds, as will be mentioned later.

After the war, the Society of Chemical Industry of Japan joined the Chemical Society of Japan, and together they resumed their activities in mid-1948 under the name of the latter group. During their annual meetings (1949–1950), we presented a series of papers on the structure of hinokitiol[216] and its nitro derivatives,[71c] the synthesis of tropolone,[71d] as well as thujaplicins.[71e] After these presentations, I decided to publish all my subsequent papers in English. However, at that time, the *Bulletin of the Chemical Society of Japan* began some publications, although at irregular intervals. In those days, there was no way of rapid publication of an article written in English in Japan.

After the war, Professor Seiji Tsuboi, General Secretary of the Japan Academy, along with Professor Majima and others, planned to resume the war-halted publication *Proceedings of the Japan Academy.* These six- to eight-page papers, written in English, were published every month and were the main source of rapid communication of prewar Japanese research to foreign countries. In the spring of 1950, at the request of Professor Majima, I sent five papers regarding research we had just presented at the annual meeting of the Chemical Society of Japan. Professor Majima assured me these papers would be published within the year.

Tropolone Chemistry Abroad. At the end of May 1950, at the library of the American Cultural Center in Sendai, I found Doering's short announcement of the first tropolone synthesis,[74] which had been published in the May 1950 issue of the *Journal of the American Chemical Society* (*JACS*). Doering obtained tropolone **51** by the oxidation of tropilidene (cycloheptatriene) with $KMnO_4$ and isolated **51** as the Cu complex in only a 1% yield.[74] The properties of tropolone that he obtained were the same as the properties our group had found.

In May 1950, I was introduced to Professor Wyman, a professor of biology at Harvard University, who was in Japan to inspect the postwar biological and biochemical research being performed. I explained our research on tropolone and told him that by the end of 1949 we had succeeded in synthesizing tropolone and thujaplicins and were now studying their chemical reactions and physical properties. Professor Wyman advised me that we should immediately submit our research to the *JACS* as short communications. He assured me that our papers would be accepted even though we were not members of the American Chemical Society. Thus, we rushed five papers to the United States in early June of that year: (1) The Synthesis of Tropolone, (2) The

Professor Majima with Nozoe during the former's visit to Tohoku University, Sendai, in 1958. Behind them, from left: Professor S. Fujise (organic chemistry), Professor S. Hakomori (analytical chemistry), Professor H. Tominaga (theoretical chemistry), and Associate Professor K. Seto (analytical chemistry). Third row from left: Associate Professor H. Azumi (physical chemistry), Drs. T. Nakajima and H. Kon (theoretical chemistry), and Associate Professor S. Hishida (organic chemistry).

Synthesis of Isomeric Thujaplicins, (3) Dipole Moments, (4) Magnetic Susceptibilities, and (5) UV and IR Spectra.

In June 1950, before we received the referee's comments on these communications from the United States, the Japan Academy unexpectedly resumed their publication of *Proceedings of the Japan Academy*, and Professor Majima forwarded our papers on to them. Three papers were thus published in the July issue of the *Proceedings*.[67–69] This is the reason the *Proceedings* of this year did not have consecutive page numbers. Then, in September of the same year, the comments on the five papers that had been submitted to *JACS* were returned to us. *JACS* was very interested in our research. However, we were told that it was difficult to publish the papers, because some references[39] were in Japanese and were not available in the United States. Furthermore, the editors of *JACS* wanted the papers to be expanded. They informed us that they would gladly accept full papers if we could make these alterations rather than using short communications. However, we abandoned this idea, because three papers about our synthesis[67–69] had already been published by that time, and further publications of our work in full papers for the United States would require too much time.

However, one of my co-workers Professor Kubo, had been very disappointed by the news of the rejection of his paper (paper 3) on the dipole study as a *JACS* communication. He decided to send this paper to *Nature*. This report was accepted and published[75] with the editor's comment, "not only in England and the U.S., it seems, that tropolone was synthesized but also in Japan by Professor Nozoe." Following this publication, we learned that in Britain in July 1950, Cook and his co-workers[76] had synthesized tropolone by almost the same method as we had used. Furthermore, Haworth and his co-workers[77] obtained 4-methyltropolone from purpurogallin. It is interesting to note that the synthesis of tropolone was being carried out in four different laboratories throughout the world at approximately the same time in 1950.

In November 1950, a symposium, "Tropolone and Allied Compounds," was organized by the Chemical Society of London.[78] During this conference, many topics were discussed, including the synthesis of tropolone and thujaplicins, studies of the physical aspects of the tropolone nucleus by means of UV and IR spectra, and X-ray analysis of the copper complex of tropolone. Our study on hinokitiol in Taiwan was briefly introduced by Professor H. Erdtman of Sweden. Previously, I had sent Professor Erdtman a hinokitiol specimen and its various derivatives through Professor Sugasawa. Thus, Professor Erdtman had clearly appreciated that his β-thujaplicin and my hinokitiol were identical. Thus, the chemistry of tropolones, along with our studies of hinokitiol in both Taiwan and Japan, rapidly gained the attention of many chemists.

Upon learning of the 1950 symposium through the announcement in *Chemistry and Industry (London)*,[78] I immediately sent a paper outlining our research in Taiwan and Japan to the chairman of the meeting, Professor J. W. Cook (at the end of 1950). Although we were in a sense competitors, Professor Cook made arrangements to publish my entire paper in *Nature*.[79] He commented in his letter that he had not realized that such extensive research had been taking place in Japan. I was impressed by his attitude and the respect that he showed us. Because this was our second paper to be published in *Nature*, our research seemed to be gaining the publicity that it had not previously attained. Consequently, in 1951–1952, several review articles on early tropolone chemistry were published.[80–85] Cook[80] and Johnson[82]

That Outpost of Empire, Australia
Produces some curious Mammalia,
The Kangaroo Rat,
The Blood-Sucking Bat
And Arthur J Birch, inter alia.
a very pleasant + profitable visit 2/10/61
AJB.
University of Manchester.

One of my most interesting tasks was to
review in 1960 about 150 papers on tropolones
for the Annual Reports of the Chemical
Society. The novelty of this work made it
a great pleasure to meet the author of it.
My "tropone" is

Arthur J Birch ANU
Australia

5/9/77 $Fe(CO)_3$

Arthur Birch, for many years a professor of organic chemistry at Manchester, England, before moving to the Australian National University, visited Sendai in 1961 and 1977. He introduced himself as "curious mammalia" by quoting this limerick written for and about him in 1960 by his friend, Sir John Cornforth. Earlier, in 1951, Birch had written a review introducing many of Nozoe's papers on tropolones.

referred to our earliest study, and Birch[84] and Pauson[85] cited in their reviews almost all of our studies until then.

Unfortunately I cannot remember the exact date, but one day in 1952, I received a letter and many small parcels from Dr. H. E. Zaugg of Abbott Laboratories in North Chicago. In the letter, he wrote, "My father was a teacher at an American missionary school in Sendai and was a friend of your parents. I am interested in your research on tropolone chemistry and have been following your papers. I believe you have had difficulty obtaining foreign scientific journals. Because I have easy access to the library at our company, I have decided to present you with my personal copies of the back issues of *Chemical Abstracts* for 10 years starting from 1940. Incidentally, when I was packing these articles, they were made into 77 packages, and I just remembered the name of a major bank in the city, No. 77 Bank. I hope these articles are of help to you." I was very excited and grateful for this show of good will from a person I had never met and felt great respect for his dedication to research. I accepted this gift with great pleasure and found the articles to be of great help to me. After this fortunate incident, I visited Dr. Zaugg's laboratory several times. I even stayed with him when I visited Chicago.

Development of Troponoid Chemistry

Following the earliest tropolone syntheses in 1950 in the four laboratories mentioned previously, other laboratories in the United States and Europe were also succeeding in various kinds of interesting tropolone syntheses. Thus, along with the development of more practical methods[86] for tropolone synthesis, troponoid chemistry was gradually being established as a new field of organic chemistry. There are so many research papers in this field, including more than 500 of ours, and I will refer to original papers only in special cases; for detailed references, extensive reviews[85,87–93] and monographs[94,95] are available. Very recently, extensive reviews on the carbocyclic π-electron system were published by Professor Toyonobu Asao and Masaji Oda[96,97] (former students of mine), Becker et al.,[98] and Zeller.[99]

Electrophilic Reactions of Tropolones. Once we succeeded in synthesizing tropolone and its alkyl and aryl derivatives, we studied the electrophilic substitution reactions of these compounds. As expected, substitutions took place preferentially at C-3, C-5, and C-7 (*see* **39**). Moreover, we found that the priority order of the electrophiles for the substitution at C-5 (the *para* position) was as follows: NO > arylazo > NO_2 > halogens > CH_2OH, NRR′.

Unlike benzenoid compounds, tropolone did not undergo the Friedel–Crafts or Gattermann–Hoesch reaction or sulfonation with concentrated and fuming sulfuric acid. However, we found that sulfonation could be achieved by heating a solid mixture of tropolones and sulfamic acid at 150–160 °C. In electrophilic reactions, tropolones have been found to be extremely susceptible to steric effects by the side chains.[87] For example, on nitrosation with nitrous acid, tropolone gave the C-5-substituted product quantitatively, but 4-methyltropolone (**63**, R = CH_3) produced a 5-nitroso compound (**64**, R = CH_3) and the rearrangement product (**65**, R = CH_3) in a ratio of 1:1, whereas hinokitiol exclusively afforded the rearranged product [65 R = $(CH_3)_2CH$]. Sulfonation of tropolone with solid sulfamic acid took place at C-5, but the 4-isopropyl homologue was sulfonated at C-7. However, 3-isopropyltropolone gave no sulfonated product, possibly because of the transformation to a nonplanar structure at high temperatures.

We found that azo-coupling reactions normally occurred at C-3 if C-5 had a substituent. Hinokitiol exclusively gave the C-5-substituted product despite the presence of a bulky group at C-4. Hinopurpurin, which was produced by heating 5-arylazohinokitiol, **43**, in ethanol or acetic acid, was shown to have the structure **44**. Apparently, hinopurpurin is derived from the cyclization between the side chains and subsequent dehydrogenation.[100] Ring closures of this kind could also be observed for 5-arylazotropolone, **66**, bearing an ethyl or a methyl substituent at C-4, although the reaction became somewhat slower because of the decreased bulkiness of the substituent. The product, quinone **67**, existed in the enolized form, **67a** in this case. Interestingly, **68** (R = CH_3), bearing an acetamido group on the side chain at C-4, produced the same product, **67**, by elimination of the acetamido group.[90]

We found a great diversity of reactivities of tropolones. As I had noticed in the research on hinokitiol in Taiwan, synthetic tropolones possess some olefinic properties; thus they yield addition–elimination products on halogenation, as well as Diels–Alder adducts.[90] For example, tropolone readily gave the trichloro derivative **69**, which became **70** by further chlorination in acetic acid, rearranged into the hydroxy acid **71**, and finally became chloranil, **72**.[90]

5-Nitroso or 5-arylazo compounds (**73**, X = O or NC_6H_5), in the form of the monoxime or phenylhydrazone (**73a**) of the tautomeric 1,2,5-tropoquinone, reacted with *o*-phenylenediamine to give quinoxaline, **74**. This compound became quinoxalotropone, **75**, on hydrolysis with a strong acid.[89]

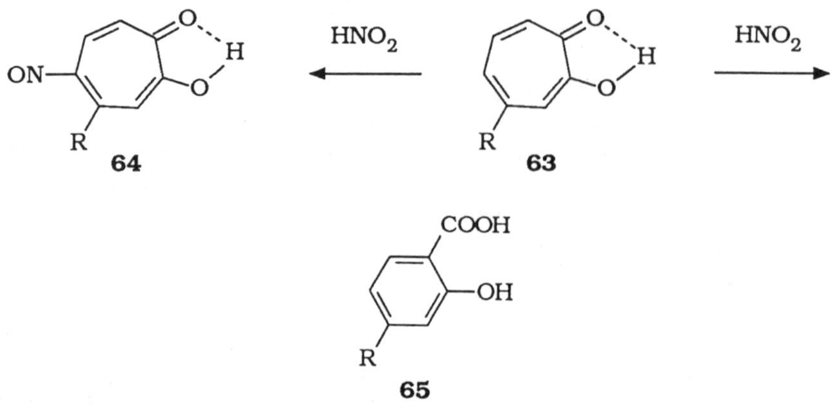

ON— R **64** ← HNO_2 — R **63** — HNO_2 →

COOH, OH, R **65**

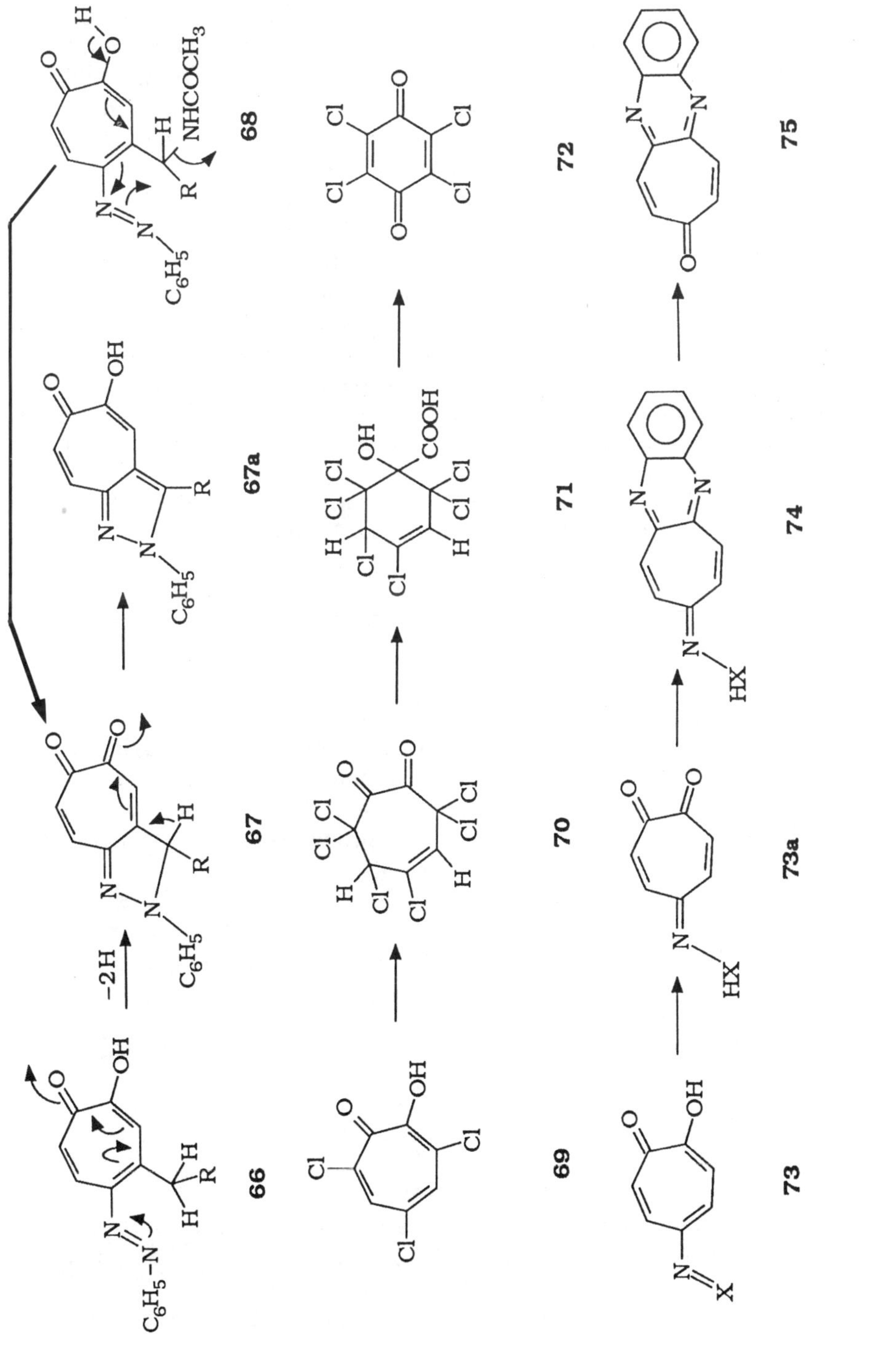

66 67 67a 68

69 70 71 72

73 73a 74 75

We carried out a careful reexamination of the perbromination of hinokitiol in aqueous acetic acid at 0–5 °C.[90] Some of the results are shown in Scheme II. 3,7-Dibromohinokitiol (**76**) gave, via the tribromodiketone **77**, a pentabromodiketone (**78**) in good yield, which in turn led to 3,5,7-tribromohinokitiol (**79**, as the *p*-toluidine salt) when heated in *tert*-butyl alcohol and then with *p*-toluidine. On the other hand, bromination of 5,7-dibromohinokitiol (**80**) yielded a tribromodiketone (**81**, as a hydrate) in good yield. The structures of these compounds were established by IR and NMR spectra. Isomerization of **81** to its tautomer (**79**), although relatively difficult, was accomplished by heating **81** in *tert*-butyl alcohol and then adding *p*-toluidine to yield the salt of **79** in good yield. The unexpected stability of **81** was attributed to the steric repulsion between the bulky vicinal substituents at C-3 and C-4. The free tribromohinokitiol was very stable once it was isolated, and the seven-membered nucleus of this compound was determined by X-ray analysis to be in a boat form.[90]

3,5-Dinitrotropolone, like the corresponding dinitrohinokitiol, was rearranged to 2,4-dinitrobenzoic acid, when heated with aqueous acetic acid, and to its ester when heated with alcohol. 3,5,7-Trinitrotropolone also yielded the 2,4,6-trinitrobenzoyl ester on heating with alcohol, whereas on heating with dilute acid or sodium bicarbonate solution, it yielded 1,3,5-trinitrobenzene.[101]

Although the olefinic character of tropolones is considerably suppressed because of their aromatic stability, tropolones turned out to be highly susceptible to oxidation with ozone or $KMnO_4$. With alkaline hydrogen peroxide, nucleus hydroxylation preferentially occurred at C-3, C-5, and C-7. This finding corroborated the previous results we had obtained with compounds A and B[39] produced by the H_2O_2 oxidation of hinokitiol, as illustrated in Scheme III.[102]

By using temperature-dependent ^{13}C NMR spectroscopy, we found that tropolone acetate undergoes a swift, degenerate rearrangement, **82a** ⇆ **82b**, at above ~–30 °C with an activation energy of ~10.8 kcal/mol.[103]

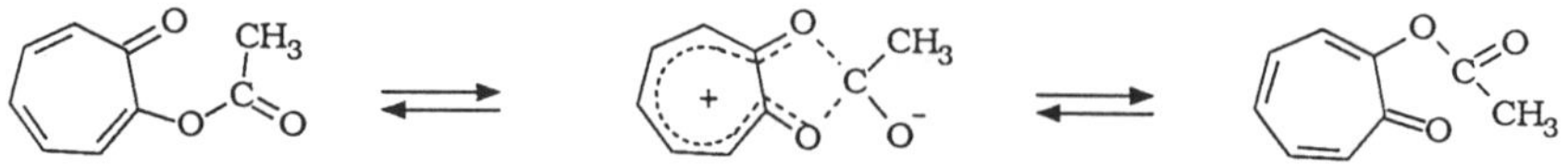

82a **82c** **82b**

Nucleophilic Reactions of Tropones and Tropylium Ions. Coincidentally in 1951, Doering,[104] H. Dauben,[105] and our group[106,107] succeeded in synthesizing tropone, **83**. Doering and his co-workers[104] considered it reasonable to call tropone "tropylium oxide" (**83a**), because the com-

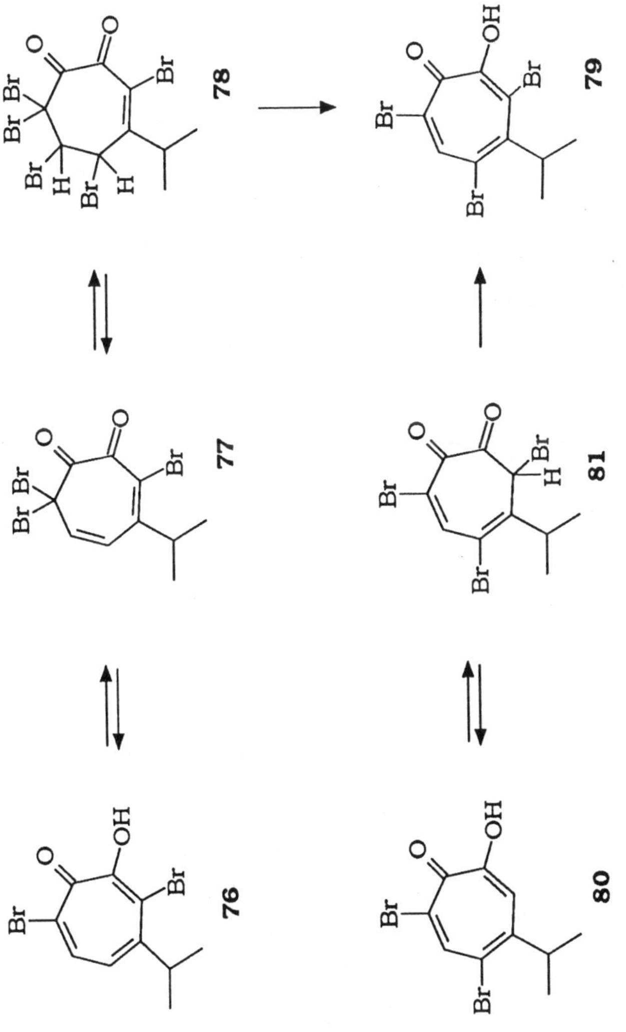

Scheme II

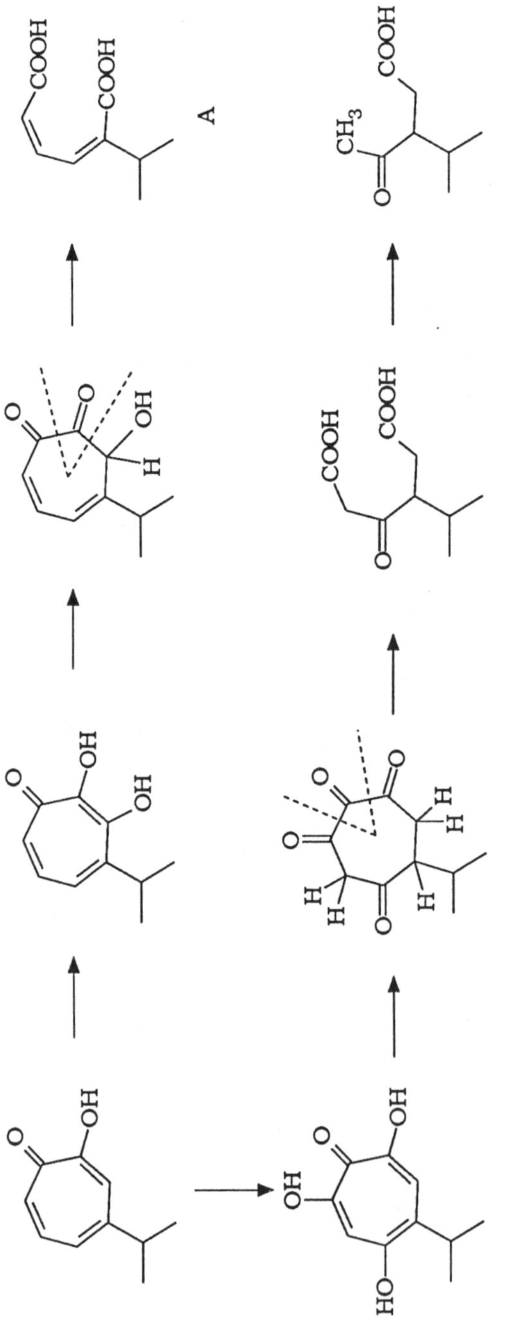

Scheme III

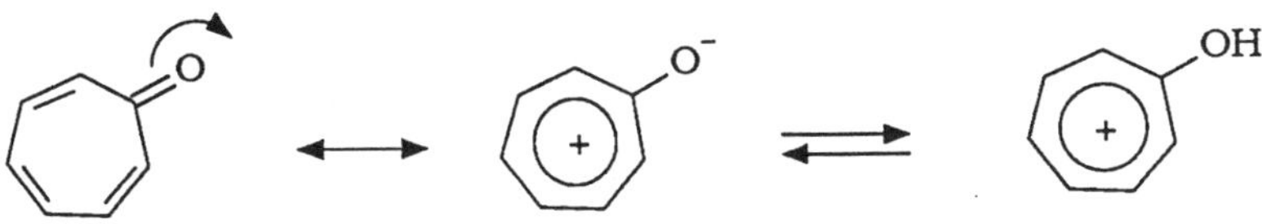

83 **83a** **84**

pound is soluble in water, the boiling point is much higher than that of benzaldehyde, and it easily produces the troponium ion, **84**, in acid. They quoted the Hückel rule—(4+2) π aromaticity—in explaining the stability of tropone for the first time. This explanation drew the attention of many organic chemists to this rule. By measuring the dipole moment of tropone, we proved that the contribution of the ionic structure, **83a**, to tropone was substantial.[108] We also proved that tropone had a double-bond character, which appeared to be considerably greater than that of tropolone, and thus tropone was more susceptible to addition–elimination reactions with halogens, as well as the Diels–Alder reaction.[109]

Tropone, **83**, produced trinitrotropolone upon treatment with concentrated nitric acid,[101] and dihydrotropone derivatives were formed by addition of a Grignard reagent or alkyl lithium[90] (Scheme IV). Moreover, tropone reacted with ketone reagents to give an oxime and a hydrazone, whereas with hydrazine hydrate or alkaline hydroxylamine, tropone yielded 2-aminotropone, **85**, almost quantitatively.[107] Because alkaline hydrolysis of **85** readily produced tropolone quantitatively, an efficient method to synthesize tropolones from tropones was established.[110] The reaction is believed to proceed[93] as follows:

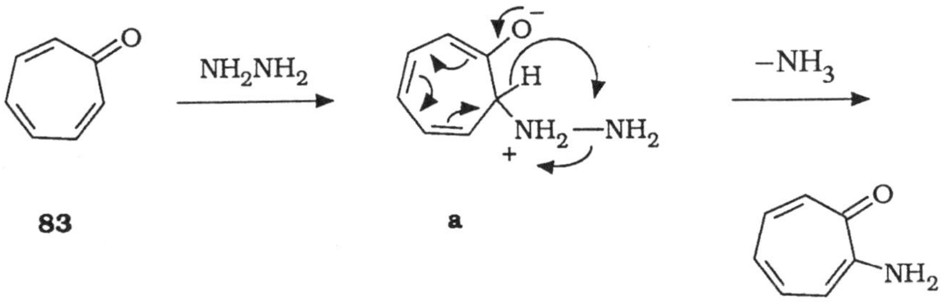

NH_2NH_2 $-NH_3$

83 **a** **85**

The groups in other countries, as well as my former co-workers (Professors Kitahara, Mukai, and Itô), reported various methods by

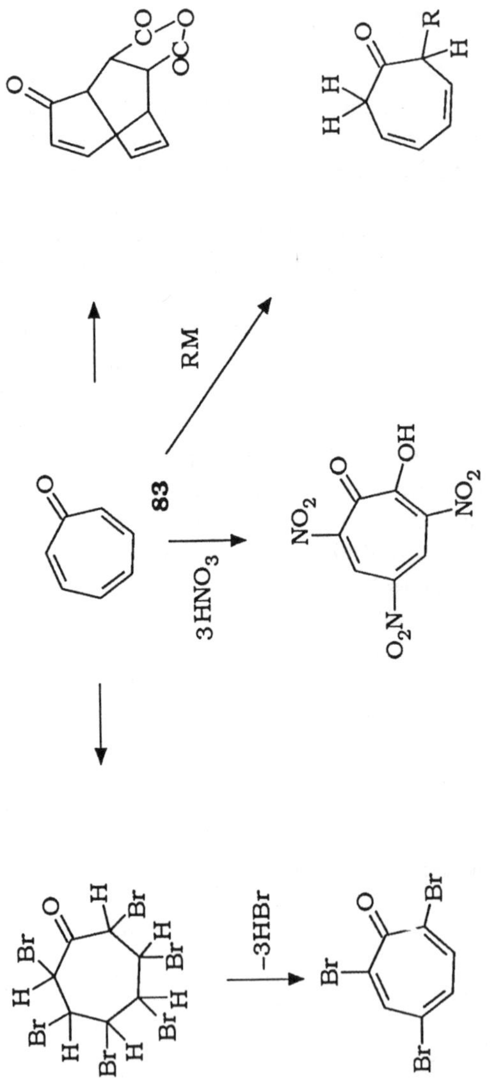

Scheme IV

which tropones were synthesized, in addition to their interesting reactivities.[93] Also, ways to demonstrate the Woodward–Hoffmann rule or to synthesize special compounds, such as the cycloaddition[93] and photocyclization products of tropones, were studied actively.[90,91,93] (Toshio Mukai was a forceful and nonbashful (in a positive sense) student who always became intensively engaged with any research topic. He had interests in the areas of photochemistry and thermochemistry while still

Tetsuo Nozoe and T. Mukai in the laboratory at Sendai, 1953.

a member of my group, and after becoming independent, he made prominent accomplishments in these fields. Recently, he also became involved in the solid-state chemistry of π-electron systems.)

Shô Itô was one of the first students to join my group at Tohoku University. He was a sociable and amiable youth. He spent 2 years as a postdoctoral fellow with Professor K. Wiesner in New Brunswick, Canada, where there were no other Japanese, and then another year with Professor R. B. Woodward at Harvard University. At Harvard, he participated in the total synthesis of chlorophyll. In addition to tropoquinone derivatives, I encouraged Itô to study thujopsene, widdrol, and some other natural products as an extension of studies I had initiated in Formosa on sesquichamaene, a constituent of the *taiwan hinoki* leaf oil.[5] After becoming independent, Shô Itô retained his interest in natural products and continued to publish important papers dealing with the syntheses of polyterpenoids, in addition to contributions in the area of nonbenzenoid "phanes". Being exceptionally fluent in English, particularly in the spoken language, he also became a president of the Organic Division of IUPAC (International Union of Pure and Applied Chemistry) and is currently a bureau member. He was the general secretary at the IUPAC 1st ISNA (International Symposium on the Chemistry of Aromatic Nonbenzenoid Compounds) International Meeting, which I organized in 1970 in Sendai, and was the chairman of the IUPAC 16th International Symposium on the Chemistry of Natural Products in Kyoto in 1988.

Reactive Troponoids. We also found that the reactive troponoid **86**, when it has a good leaving group, such as a halogen, methoxy group, or tosyloxy group, at C-2, is highly reactive toward nucleophiles. This reactivity was quite different from that of benzenoids. For example, **86** is susceptible to normal substitution at C-2, as well as the ciné substitution (substitution at a different position) at C-7,[91,94,111] depending on X and the reaction conditions. The ring contraction of **86** produces benzoyl, salicylaldehyde, and *m*-hydroxybenzaldehyde derivatives,[91,98] depending upon which position the nucleophile attacks (Scheme V). Thus, as an extension of this substitution reaction, many kinds of troponoids have been synthesized by using various nucleophiles, such as alkoxides, alkalies, sulfides, mercaptides, amines, hydrazines, organometallic compounds, and active methylene compounds.[91–95]

We also studied the mechanism of the Grignard reactions. The facts that two equivalents of the reagent were always required for the completion of the Grignard reaction of the reactive troponoids and that a D atom was incorporated into the position of the methoxy group upon quenching the intermediate complex with D_2O proved the ciné-substitution mechanism[112] previously speculated by Haworth[111] for the

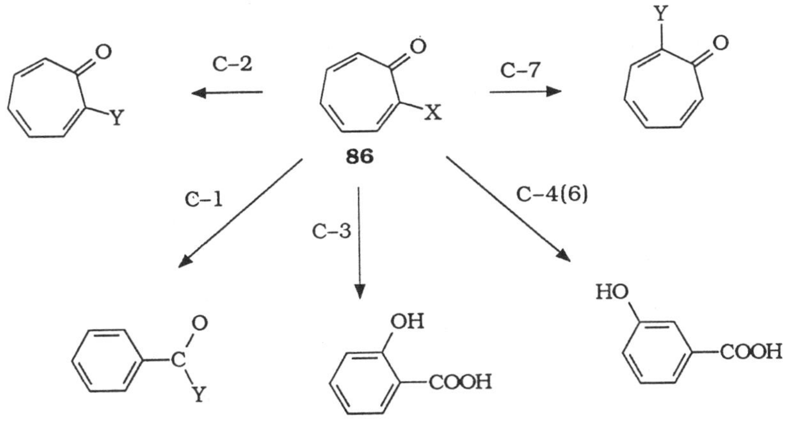

Scheme V

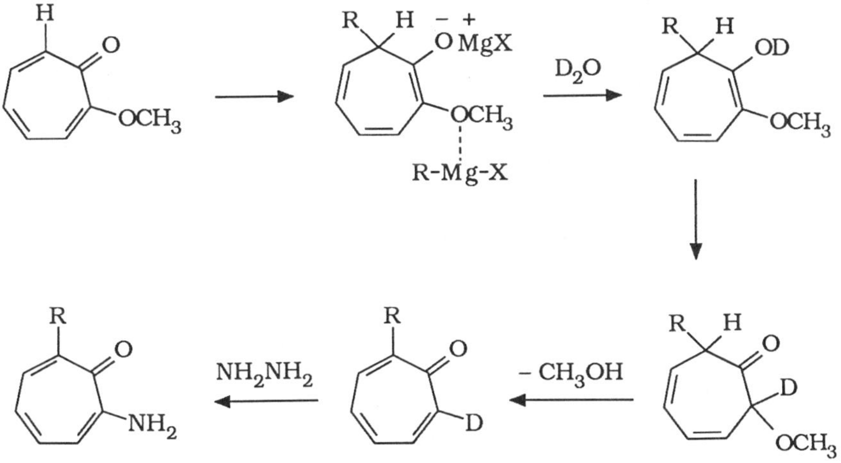

Scheme VI

ciné reactions (Scheme VI). Later, a systematic study of the kinetics of the reactions of C-2-substituted tropones was performed by Pietra and his co-workers in Italy.[91]

Three isomeric methyl ethers (3-, 5-, and 7-bromo-2-methoxytropones), **87–89**, were readily prepared from 3- and 5-bromotropolones. The positions of the nucleophilic attack and the type of reaction products from these ethers were quite different and very interesting. Noteworthy examples will be described in a later section (*see* Cyclohepta[*b*][1,4]benzoxazine).

Tropylium Ion. By heating tropilidene (cycloheptatriene), **90**, with bromine in acetic acid, tropylium salt, **91**, was obtained by Doering et al.[113]

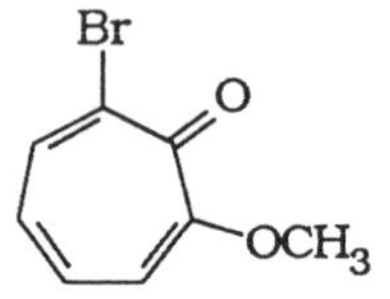

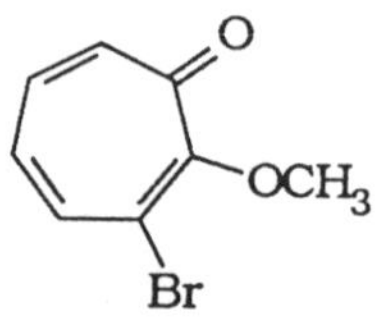

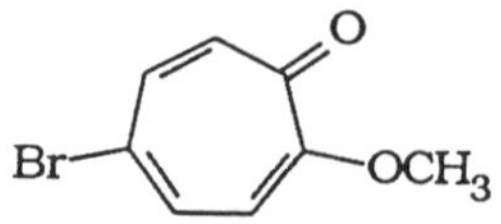

87 88 89

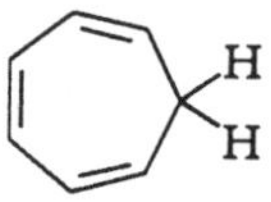

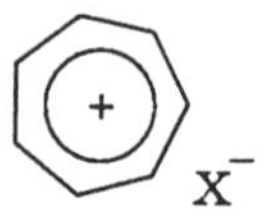

90 91

After this tropylium ion was established as the parent troponoid compound, research on the synthesis and reactions of this cation intensified in many countries.[91,114] Because tropyl methyl ether 92 is easily soluble

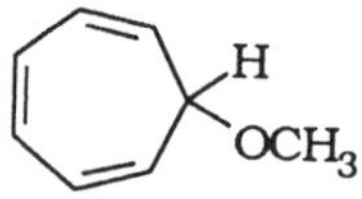

92

in common organic solvents and reacts as a cation, 91, in the presence of a trace of acid, we used 92 as a tropylation reagent.[114,115] Ditropyl ether, 93, which is obtained by adding dilute alkali to 91, produces an

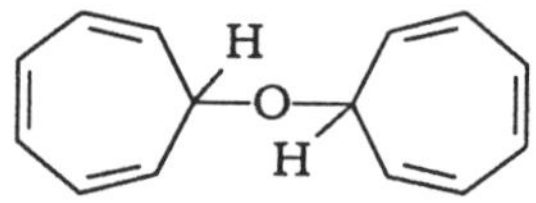

93

equimolecular quantity of tropilidene and tropone[116] because of a disproportionation reaction by the action of a trace of acid. This reaction was also demonstrated in several other laboratories abroad at around the same time.[91,99]

C-7-substituted tropilidenes, **94a**, which were prepared by the reaction of **91** or **92** with various kinds of nucleophilic reagents, became a mixture of isomeric compounds, **94a–94d**, by a 1,5-hydrogen shift, when heated at temperatures higher than 100 °C.[117] Because the mixture **94b–94d** does not revert to cation **91** by the action of acid, it can be converted first to the substituted tropylium ions (**95**) by dehydrogenation with trityl (triphenylmethyl) salt in acetic acid, and then tropones (**96** and its isomers), 2-aminotropones (**97** and its isomers), and tropolone derivatives (**98** and its isomers) can be obtained. Thus, for example, we

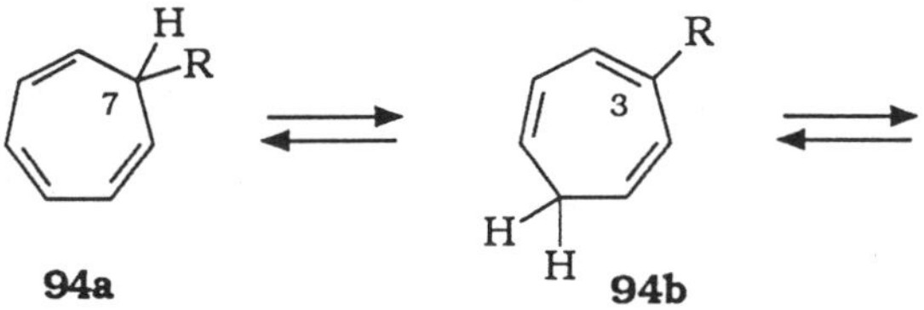

94a **94b**

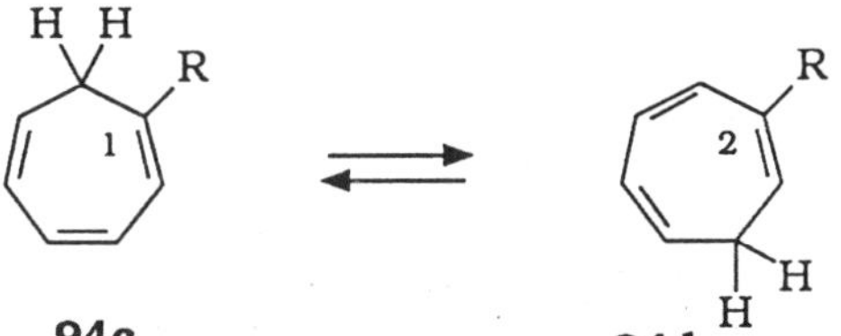

94c **94d**

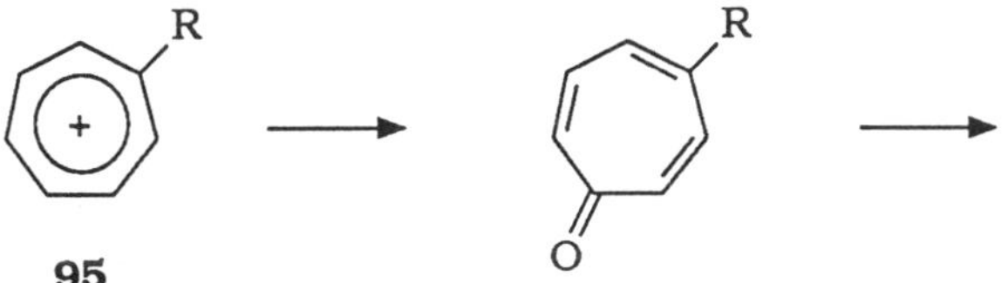

95 **96**

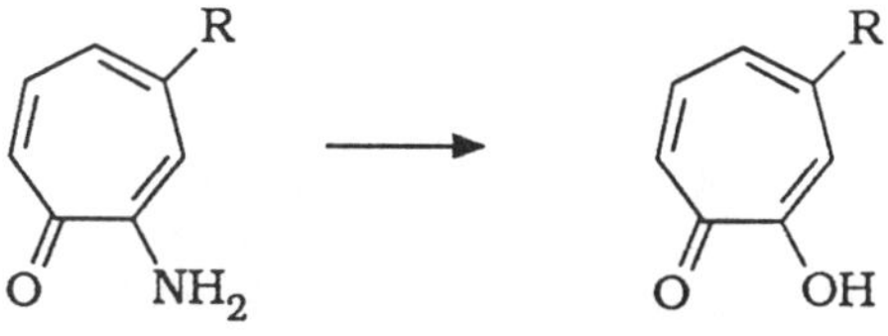

97 **98**

succeeded in obtaining aryltropones (**101** and its isomers), aminotropones, and tropolones (**102** and its isomers, X = NH_2 or OH)[118] via **99** and **100**.

Fine Structure of Troponoid Nucleus. As I described earlier, I organized a large cooperative group at Tohoku University.[119] I invited key figures in the field of physical chemistry in Japan to join the group and gradually widened the research activities to include the following fields at various universities: dipole moment, electron diffraction, and quadrupole resonance (M. Kubo, M. Kimura, and Y. Kurita at Nagoya); UV and IR spectra (T. Shimanouchi, M. Tsuboi, and K. Kuratani at Tokyo and S. Kinumaki and Y. Ikegami at Tohoku); X-ray analysis (I. Nitta, T. Watanabe, and Y. Sasada at Osaka); magnetic susceptibility (G. Hazato and J. Maruha at Tohoku); near-UV and Raman spectra (S. Imanishi and M. Ito at Kyushu); ionization potential (K. Higashi at

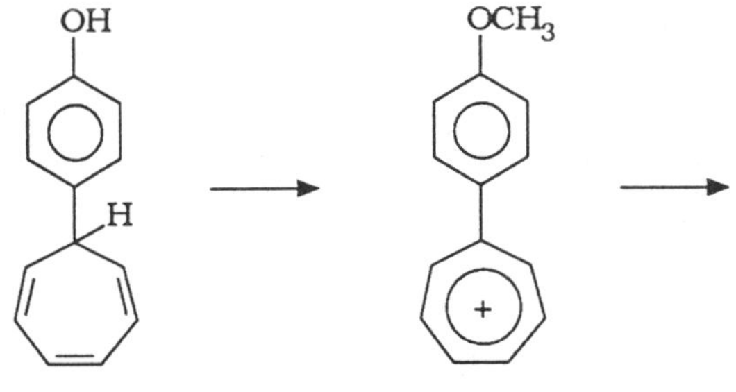

99 100

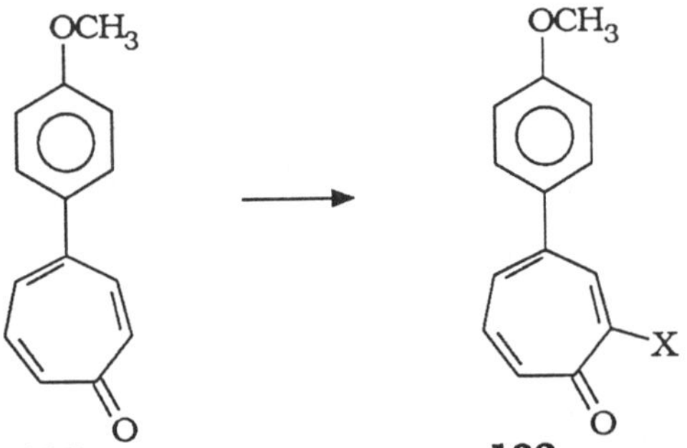

101 102

Hokkaido); mass spectra (C. Djerassi and J. M. Wilson at Stanford); polarography (S. L. Santavý at Oulomoucz and N. Tanaka and T. Takamura at Tohoku); dissociation constants (N. Yui at Tohoku); metallic complexes (U. Uemura at Tokyo Institute of Technology, Y. Oka and S. Matsuo at Tohoku, and S. Yamada at Osaka); and molecular orbital calculation (M. Kubo at Nagoya and H. Kon at Tohoku).

In the early days of troponoid research, we, as well as other scientists abroad,[45,48] assumed that the tropolone ring existed as a resonance hybrid of two intramolecularly hydrogen-bonded structures (**33a** and **33b** and **51a** and **51b**). However, UV and IR spectroscopic methods showed that the tropolone ring is a high-velocity tautomeric system of two structures, each of which is stabilized by the resonance of ionic structures.[48,120–122] Thus, Dewar's hypothesis[46] proved to be correct.

Because of our joint research, we were able to confirm that tropolone exists in a monomeric, intramolecularly hydrogen-bonded form[121,122] in a dilute solution but as a dimer in the solid state. Refined X-ray analyses made it possible to discern that solid tropolone had dimeric bifurcated hydrogen bondings, **103**,[123] whereas the lengths of the two C–O bonds of the tropolonate[124] (**104**) and troponium[125] (**105**) ions were the same. Also, X-ray analysis of the two bromomethoxytropones, **87** and **88**, whose reactivities differ considerably, allowed us to

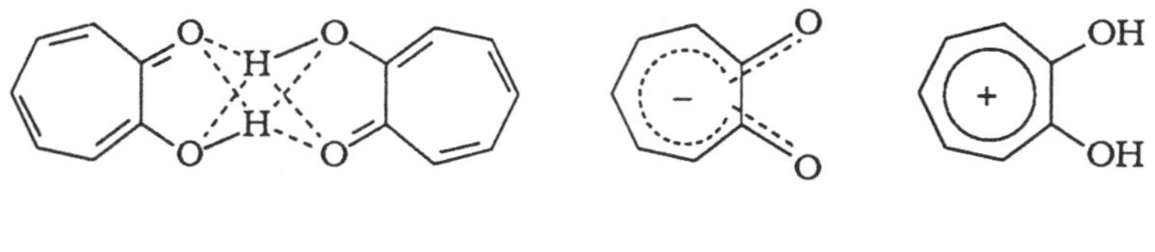

103 **104** **105**

ascertain that, especially in 3-bromo-2-methoxytropone (**88**), the methoxy group on C-2 was almost perpendicular to the plane of the seven-membered ring.[126]

Through physical measurements and the earlier mentioned chemical reactivities, we became convinced that the monocyclic troponoid ring was a novel type of aromatic system, which, to some extent, had a unique and stabilizing six-π electronic structure. In the 1970s, some researchers abroad reported that troponoids should be regarded as cyclic polyenones rather than as aromatic compounds, on the basis of NMR or X-ray diffraction analyses of some troponoid compounds.[91,93] However, we believe that such troponoids as those having a monocyclic system or an aryl side chain, as well as tricyclic colchicine derivatives, possess, at least at the moment of their chemical reaction, an aromatic character contributed by a six-π-electron tropylium system. Otherwise, we cannot explain a number of the unique properties revealed by our numerous experiments.

S and N Analogues of Tropones, Heptafulvenes, and Quinarenes. We attempted the synthesis of compounds that bear a N or S atom(s) instead of an O atom(s) in the troponoid nucleus. The aforementioned 2-aminotropone (**85**) is also produced from 2-methoxytropone and ammonia, and the site that is susceptible to electrophilic substitution is similar to that of tropolone.[127] The substitution product **106** gives the corresponding tropolone **106a** by alkaline saponification without rear-

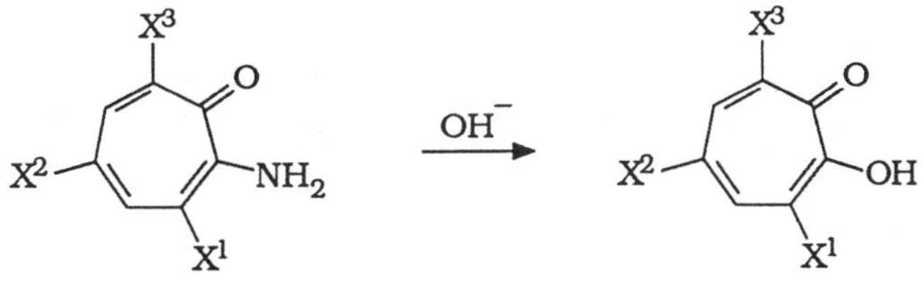

X^3, O, X^2, NH_2, X^1 —OH^-→ X^3, O, X^2, OH, X^1

106 **106a**

rangement, even if it contains substituted halogens. Hence, by this method, the position(s) of the substituent(s) in **106** could be determined. Furthermore, by the catalytic reduction (over Pd–C)[128] of tropoquinone trioxime (107), which is readily obtainable from 5-nitrosotropolone and hydroxylamine, we were able to obtain 2,5-diaminotroponimine (108). *N,N'*-Dibenzoylaminotroponimine (109, R = C_6H_5CO), which is

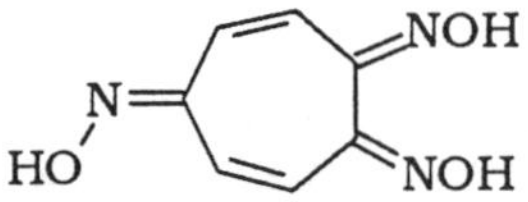

107

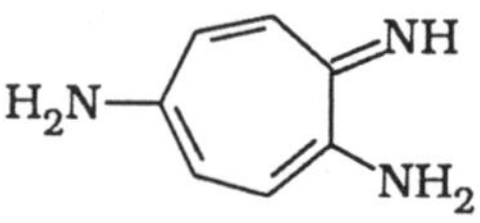

108

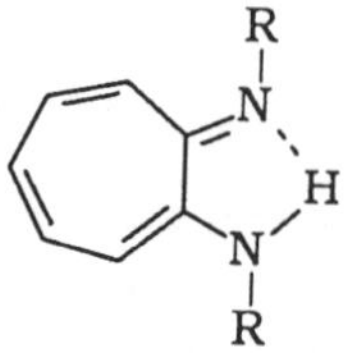

109

obtained by the Schotten–Baumann benzoylation of 1,3-diazaazulene, was converted to the 2-phenyl derivative of 1,3-diazaazulene (*vide infra*).[129] 2-Arylaminotroponimines easily produce metal complexes and were studied actively by Benson et al.[91] We also obtained mercaptotropone (110) by reacting KSH with the reactive troponoid.[130] This com-

110a

110b

pound is considered to exist as tropothione (110a) and, by acylation or alkylation, becomes a S-substituted tropone, 111.[131] Compound 110 is

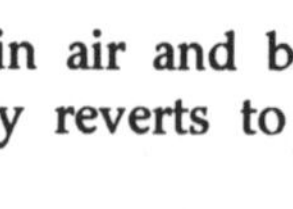

111

very susceptible to oxidation in air and becomes a disulfide, but on reduction, the disulfide readily reverts to 110. We also obtained 2-aminotropothione (112).[132]

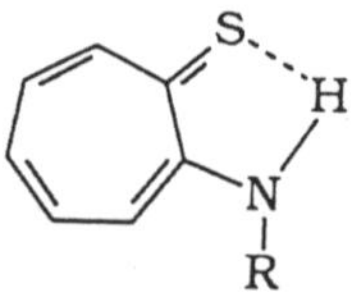

112

Heptafulvenes, 114, methylene analogues of tropone, intrigued many chemists. Theoretical calculations concerning these compounds were performed in various countries. However, the parent compound, 114 (X = Y = H), was too unstable to be isolated at that time.[133] We considered that this system would be stabilized by introducing two electron-withdrawing groups in 8,8′-positions, and we were able to synthesize 114 (X, Y = CN or COOR) from the tropyl compounds 113 and 115, which were obtained by reacting the tropylium ion with an active

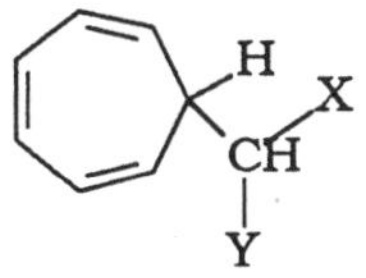

113

114 114a

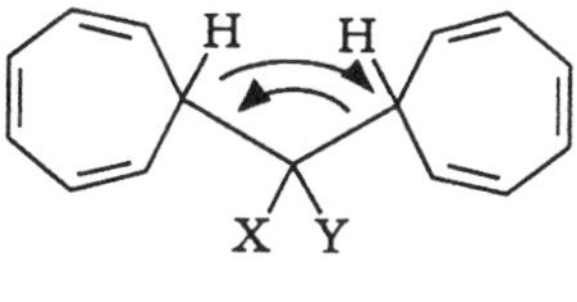

115

methylene compound.[134] In particular, when the compound had two cyano groups, 114 (X = Y = CN) was obtained as very stable orange crystals. The dipole moment measurement ascertained that the contribution of the six-π-electron electronic structure is substantial in this compound, as in tropone.[135]

Our next goal was to synthesize compounds such as 116–118, which have an inserted, cross-conjugated quinonoid system in the tropone or heptafulvene system. We first isomerized the aforementioned *p*-tropylphenol (99) and its analogues and succeeded in synthesizing compounds 116 and 117 by hydride abstraction with trityl salt. We referred to 116 and 117 as benzoquinonetropides.[136–138] At the same

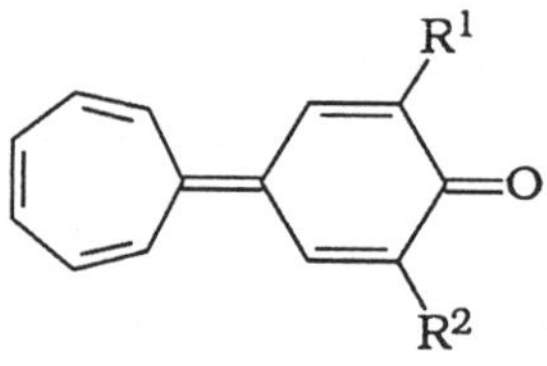

R1 X O O R2 O

116 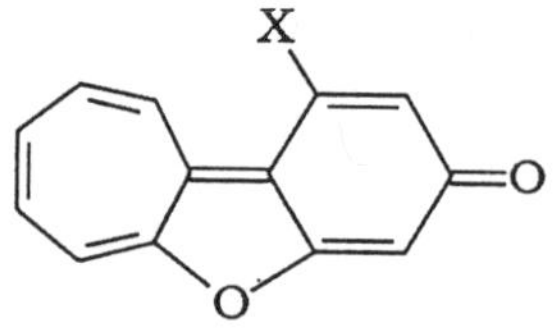 **117**

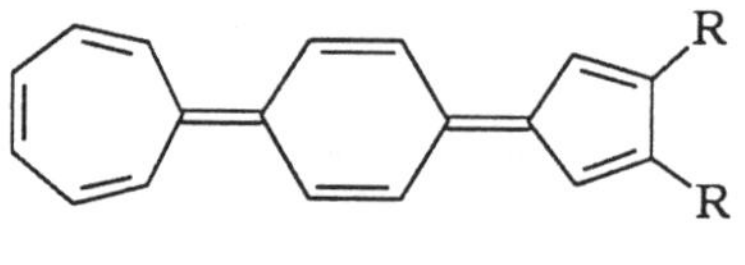

R R

118

time, Jutz (in 1964) and Bladon (in 1966) also performed similar studies.[91] We then tried to synthesize the compounds **118** and suggested that compounds of this type, which possess a cross-conjugated benzoquinonoid system, would be more generally called quinarenes.[139,140] This research was completed later,[140,141] after my retirement from Tohoku University in 1966.

Tropylium Compounds Fused with Heterocyclic Materials. To examine further the electrophilicity of the reactive troponoid **86**, we studied its reactions with guanidine, thiourea, malonic ester, and acetoacetate. Fortunately, we succeeded in synthesizing 1,3-diazaazulene (**119**, X = SH or NH_2),[142] 1-azaazulene (**120**),[143] cyclohepta[*b*]thiophene (**121**, X = NH),[144] and cyclohepta[*b*]furan-2-one or imine (**122**, X = O or NH)[145] in one

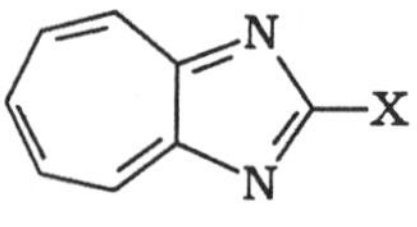

N X N R X N

119 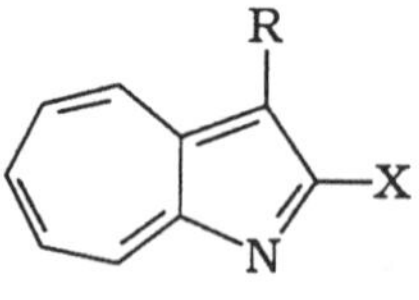 **120**

R X S R X O

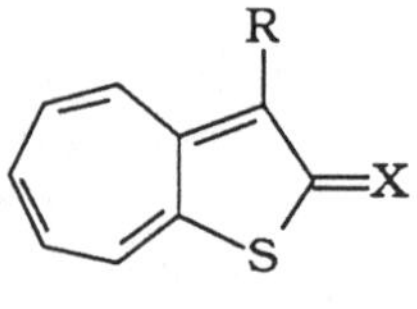

121 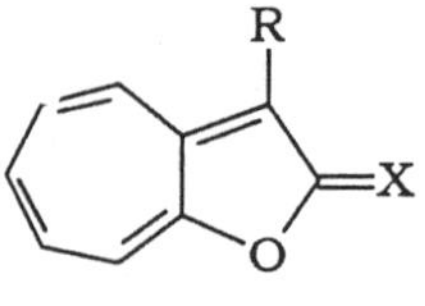 **122**

step and in very high yields. We found that 2-hydroxy and 2-amino substituents in **119** and **120** usually exist in the keto and imino forms, respectively.

The reactive troponoids annulated with a benzene or heteroaromatic nucleus are usually very stable and do not produce these heterocyclic products at all. However, just as with the reactive monocyclic troponoids, tricyclic colchicine (**123**) produces **124** and the

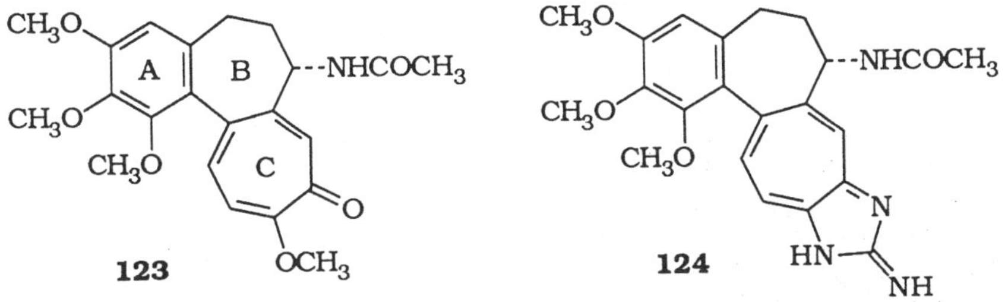

123 **124**

related compounds in a high yield.[146a] By applying the traditional heteroring closure reaction on the troponoid ring, we also obtained many kinds of heterocyclic compounds (Chart II).

Novel Synthesis of Trisubstituted Azulenes. In 1953, as a further extension of the nucleophilic reactions mentioned earlier, we studied the reaction of 2-chlorotropone with ethyl cyanoacetate (**125c**, R =

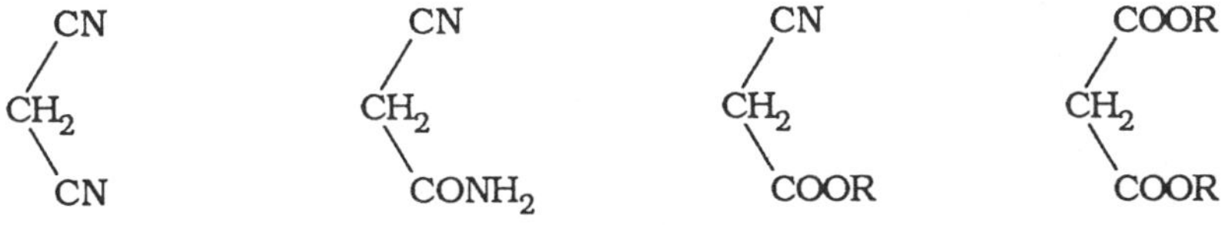

125a **125b** **125c** **125d**

C_2H_5) and expected to obtain the 3-cyano compound **122** (X = O, R = CN; $C_{10}H_5O_2N$). However, we did not obtain this result; instead, we obtained a 70% yield of orange prisms A ($C_{16}H_{17}O_4N$) and a small amount of orange needles B ($C_{14}H_{11}O_3N$). Judging from their composition, we could not assign any reasonable structures to these substances and found ourselves at a standstill for a while. Fortunately, a Baird IR spectrometer was introduced 1 year later at the analytical center of Tokyo University. By using this instrument, we learned that compound A possessed a conjugated ester carbonyl group and amino groups, whereas compound B contained a conjugated cyano and ester carbonyl groups and acidic hydroxyl groups. We had never imagined that such a

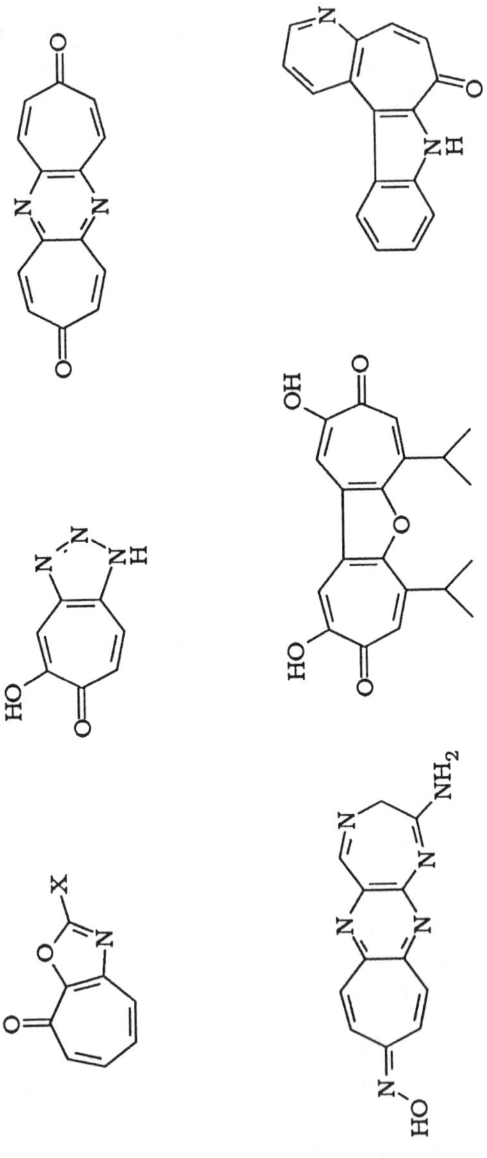

Chart II

wide variety of functional groups could be determined instantly, and we were quite impressed by this modern analytical instrument.

To explain the compositions of these two compounds and their high stability, we first assumed that they were naphthalene derivatives that could have been formed by rearrangement during the reaction. We applied the Sandmeyer reaction to compound A. When we added isoamyl nitrite to the orange A–HCl salt, the color of the solution changed from blue to violet and finally to red. The composition of the red needles obtained by this reaction was $C_{16}H_{15}O_4Cl$, as expected. However, when A was deaminated by the Griess reaction, it again turned to red crystals. Then, the two ester groups of A were hydrolyzed and decarboxylated by heating, and a compound with the same composition as naphthylamine was obtained in the form of red needles. The *N*-acetyl derivative of of the red needles formed spectacular violet crystals.

Because these compounds dramatically changed color[146b] as a result of a minor change in the structure, I considered that they must be azulene derivatives. To prove this speculation, the dicarboxylic acid obtained by saponification of the deamination product of A was heated under reduced pressure. We were excited to find spectacular blue crystals of azulene (126) emerging by sublimation. On the basis of these

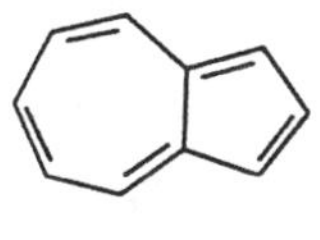

126

experiments, we learned that A was diethyl 2-aminoazulene-1,3-dicarboxylate (127, $X^1 = X^3 = COOC_2H_5$, $X^2 = NH_2$), and B was ethyl

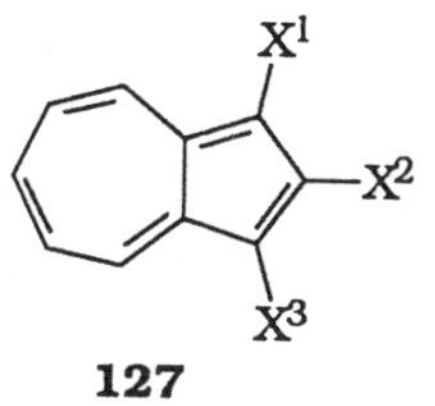

X^1 X^2 X^3

127

3-cyano-2-hydroxyazulene-1-carboxylate (127, $X^1 = COOC_2H_5$, $X^2 = OH$, $X^3 = CN$).[147,148]

It was a pleasant surprise that we were able to obtain polysubstituted azulenes, 127, in one step and in excellent yield, because we had not been able to obtain such compounds by any of the more-traditional methods. A little later, we learned that Ziegler and Hafner[149] had discovered a convenient method of general azulene synthesis by the condensation of cyclopentadiene with Zincke's salt, and this method was further developed by Hafner et al.[150] It is interesting that these two general methods of base-catalyzed azulene synthesis were discovered independently and simultaneously in 1955, and both methods give excellent yields in a very simple manner without involving any dehydrogenation steps.

By further examination using our method, we obtained four azulenes altogether, and either A or C (**127**, $X^1 = X^3 = CN$, $X^2 = OH$) was isolated in more than 70% yield, depending upon the amount and kind of base used as the catalyst. When 2-chlorotropone was mixed with 2 mol of malononitrile (**125a**), 2-amino-1,3-dicyanoazulene (**127**, $X^1 = X^3 = CN$, $X^2 = NH_2$) was precipitated almost quantitatively. Initially, we assumed that the first intermediate of this reaction was the substituted heptafulvene **114**.[148] However, upon further study, we found that the real intermediates in this reaction were in fact cyclohepta[*b*]furan-2-one derivatives (**122**), which are, in some cases, too reactive to be isolated. The intermediate instantly reacted with a second active methylene compound (AMC) to form azulene, **127**. To elucidate the reaction mechanism of this unprecedented azulene formation reaction, we isolated various first intermediates A (**122**) and then treated them with four kinds of AMCs, namely **125a**, cyanoacetamide (**125b**), **125c**, and diethyl malonate (**125d**, $R = C_2H_5$), in the presence of alkoxide as catalyst. In this way, we obtained a variety of trisubstituted azulene, usually in more than 95% yield.[151–153]

The results of these reactions are shown in Table II. In the table, the structures of azulenes, **127**, are abbreviated by the presentation of the three substituents at positions 1, 2, and 3. The minus sign (–) indicates no azulene formation under the reaction conditions used. Only the reaction of **122a** with **125c** (in Table II) gave products in which two functional groups ($CONH_2$ and NH_2) originated from the first reagent (AMC^1). A mixture of two azulenes was produced usually when **125c**

$COCH_3$–CH_2–$COOR$ **125e**

COC_6H_5–CH_2–$COOR$ **125f**

$COCH_2COOR$–CH_2–$COOR$ **125g**

was used as AMC^2. However, as mentioned earlier, the reaction of 2-chlorotropone with 2 equivalents of **125c** yielded four kinds of azulenes, apparently because of the formation of an equilibrated mixture of two intermediates **122c** and **122d** in the alkaline solution containing an alkoxide. Particularly notable among the results shown in Table II is the fact that free 2-methyl-, 2-phenyl-, and 2-alkoxycarbonylmethylazulene-1-carboxylic acids were obtained in good yields when **125c** or **125d** was reacted with A (**122f–122h**) having an acyl group at C-3. Moreover, azulenopyridines (**128a** and **128b**)[154] and azulenopyrimidine (**130**)[155] were obtained directly in very high yield, as shown in Scheme VII.

On the basis of these experiments, we proposed the reaction pathway involving the four intermediates (**A–D**) illustrated in Scheme VIII. Furthermore, we learned that two kinds of reaction pathways exist (paths a and b) in the fourth step (**C** → **D**) of this extremely complex reaction.[151, 156]

Our trisubstituted azulenes (**127**, $X^1 = X^2 =$ COOR, $X^3 = NH_2$ or OH) were so versatile that we obtained not only a variety of mono-, di-, tri-, and tetrasubstituted azulene derivatives but also numerous other azulenic compounds annulated with various heterocyclic groups, as shown in Chart III. Thus, by using our method, we synthesized a wide variety of colorful azulenes, which were usually very stable compounds.

The results described in this account were presented[156] at the First Japan–U.S. Science Seminar organized by Professor J. D. Roberts and myself in April 1965 and held in Kyoto, Japan. In attendance at this meeting were Professors J. D. Roberts, P. Bartlett, G. S. Hammond, S. Winstein, and R. Breslow from the United States and Professors

Attending members of the first Japan–U.S. seminar in physical organic chemistry pose in front of the old Kyotokaikan in Kyoto, April 1965. From left: Professors K. Maeda, G. S. Hammond, P. Bartlett, T. Hayashi, Dr. Inoue, Professor W. Tagaki, S. Oae, S. Winstein, Tetsuo Nozoe, H. Sakurai, and John D. Roberts. (Photo courtesy of H. Yukawa.)

Table II. Azulenes Produced from Intermediate A and Active Methylene Compounds

		Substituents of Product or Product with Second AMC[a]					
				125c		**125d**	
First AMC	*Intermediate A (R, X)*[b]	**125a**	**125b**	*First*[c]	*Second*[c]	*First*[c]	*Second*[c]
125a	**122a** (CN, NH)	CN[d] NH_2 CN	—[e]	$CONH_2$[d] NH_2[d] CN		—	
125b	**122b** ($CONH_2$, NH)	$CONH_2$[d] NH_2 CN	$CONH_2$[d] NH_2 $CONH_2$	$CONH_2$[d] OH CN	$CONH_2$[d] NH_2 COOR	—	
125c	**122c** (COOR, NH)	COOR[d] NH_2 CN	CN[d] NH_2 $CONH_2$	COOR[d] NH_2 COOR	COOR[d] OH CN	—	
	122d (CN, O)	CN[d] NH_2 CN		CN[d] NH_2 COOR	CN[d] OH CN	—	
125d	**122e** (COOR, O)	COOR[d] NH_2 CN	COOR[d] NH_2 $CONH_2$	COOR[d] OH CN	COOR[d] NH_2 COOR	COOR[d] OH COOR	
125e	**122f** ($COCH_3$, O)	$COCH_3$[d] NH_2 CN	$COCH_3$[d] NH_2 $CONH_2$	COOH[d] CH_3[d] CN	$COCH_3$[d] NH_2 COOR	COOH[d] CH_3[d] COOR	$COCH_3$[d] OH COOR

125f	**122g** (COC_6H_5, O)	COC_6H_5[d] NH_2 CN		COOH[d] C_6H_5[d] CN	COOH[d] C_6H_5[d] COOR
125g	**122h** ($COCH_2COOR$, O)	**128a**	**128b**	COOH[d] CH_2COOR[d] CN	COOH[d] CH_2COOR[d] COOR

[a]Substituents are given in the order X^1, X^2, and X^3. The product with the second AMC has the general structure:

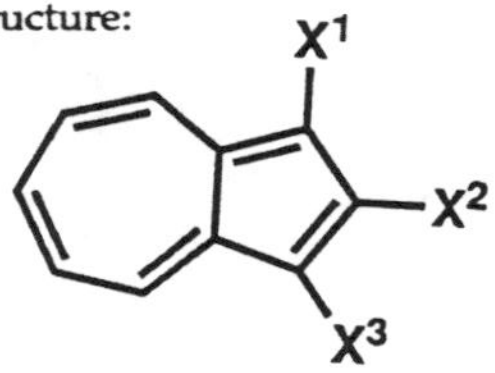

[b] Intermediate A has the general structure:

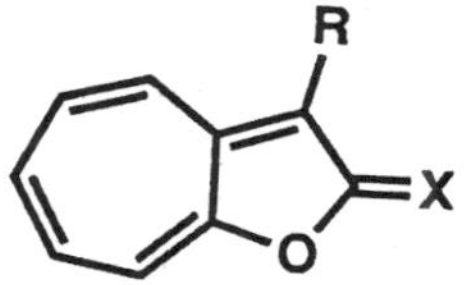

[c] If **125c** and **125d** are used as second AMC, usually a mixture of two azulenes is produced (by the difference of ring-closing functional groups and/or different reaction paths).

[d] These groups originate from the first intermediate A. The other groups are derived from the second AMC.

[e] — indicates that no azulene formation occurred under the reaction conditions.

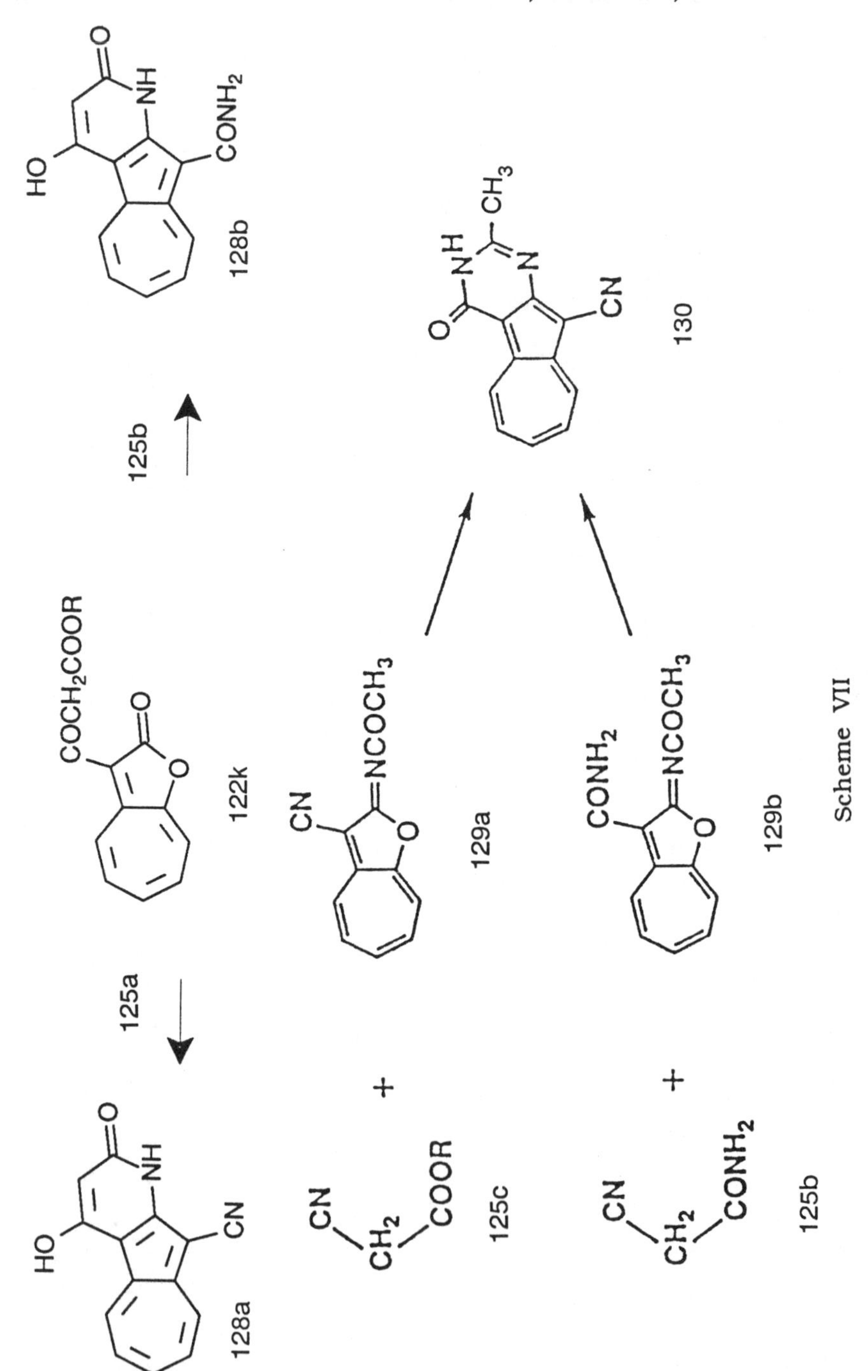
HO
NH
O
CONH2
128b
125b
CH3
NH
N
O
CN
130
COCH2COOR
O
O
122k
125a
CN
=NCOCH3
O
129a
CONH2
=NCOCH3
O
129b
HO
NH
O
CN
128a
CN
CH2
COOR
125c
+
CN
CH2
CONH2
125b
+

Scheme VII

Scheme VIII

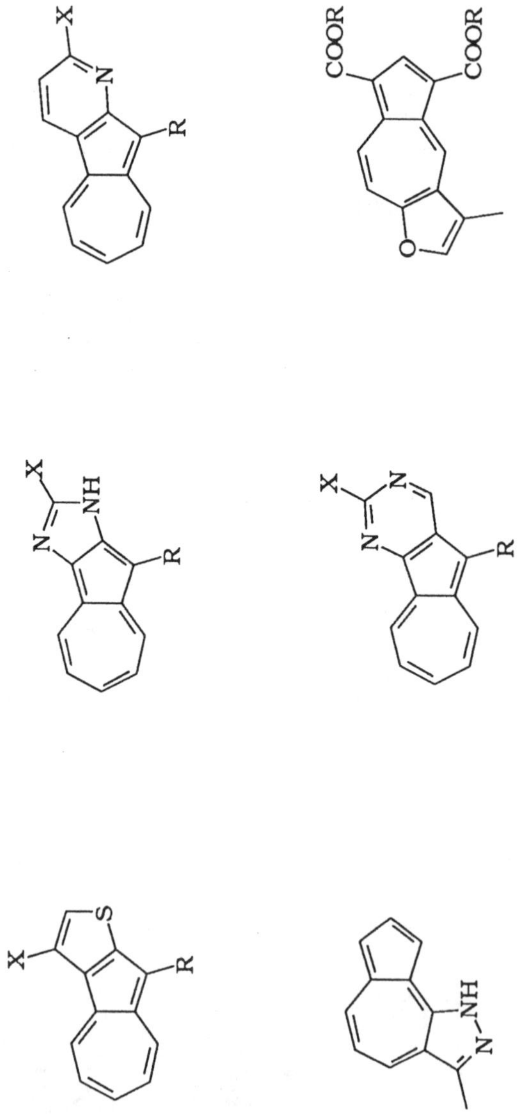

Chart III

During a recess at the Japan–U.S. seminar. From left: H. Sakurai, S. Winstein, Tetsuo Nozoe, Yoshio Kitahara, G. S. Hammond, and S. Oae.

In the garden of Katsura Imperial Villa in Kyoto, April 1965. From left: Dr. Inoue, Yoshio Kitahara, Kyoko Nozoe, G. S. Hammond, S. Winstein, Dr. Arvey, Mrs. Sylvia Winstein, Mrs. Hammond, Mrs. Bartlett, and Tetsuo Nozoe. (Photo courtesy of H. Yukawa.)

At the bus stop, ready for sightseeing in Kyoto and its environs. H. Yukawa of Osaka University, Yoshio Kitahara of Tohoku University, and Tetsuo Nozoe. (Photo courtesy of H. Yukawa.)

O. Simamura, T. Hayashi, Y. Yukawa, S. Oae, and myself, as well as eight Japanese assistant professors acting as observers. The 5-day-long seminar and often heated (sometimes stormy) discussions, carried out according to the American style of debate, was an extremely valuable and enjoyable experience.

I retired from Tohoku University in March 1966. Professor Majima reached his 88th birthday in 1961, but he passed away in August 1962. A great number of his students (first and second generations) became professors and actively conducted research in many universities and institutions in Japan and abroad. In 1926, he established Nippon Kagaku Kenkyukai (Japan Science Research Association) to publish "complete chemical abstracts of Japan" because of the extreme difficulty of obtaining chemical literature published in Japan (*see* Majima Family Tree on pp 107–109).

Direct Contact with Chemists Abroad. It was before 1935, during my days as an assistant professor at Taihoku Imperial University, that I first wished to go abroad to study natural products chemistry at the Eidgenossische Technische Hochschule (ETH) in Zurich under the guidance of Professor Ruzicka. This institute was then considered a mecca for the study of natural products chemistry.

John D. Roberts contributed his autograph and the accompanying sketch during the first Japan—U.S. Seminar, Kyoto, 1965. He complimented Nozoe in English and Japanese. The characters on the stick figure read "Japanese Chemists". Roberts learned them from Y. Kitahara.

At this time, the position of a full professorship of organic chemistry was still vacant at Taihoku Imperial University. To fill the vacancy until another suitable professor was found, Professor Kafuku gave weekly lectures on special topics in organic chemistry. As for me, I presented lectures on organic chemistry to students of both the chemistry and agricultural chemistry departments. So, obviously, if I left, no one would be available to give the organic chemistry lectures or to assist the students in their undergraduate research. Therefore, I temporarily gave up the notion of studying abroad. After I was promoted to a full professorship in 1937, I again contemplated this notion. However, the onset of war put a sudden stop to any further ideas.

When I returned to Japan in 1948, I learned that many scientists overseas were doing the same research that I was doing. Clearly, we had to contact these scientists to discuss our mutual research interests

Impressions of the First Japan-U.S. Science Seminar in Physical-Organic Chemis

A beautiful delirium
Of σ', deuterium,
Cars smaller than French Dauphines,
And colors gay from lophines
(Subtle purples, blues, and reds);
Quick warnings not to bump our heads
No time to be a golfer,
But lots of news on sulfur.
No snow that one can ski on,
But lovely girls from Gion.
Rings semi-anti-aromatic,
And taxi doors all automatic.
Free radicals and ions strange,
And signals broadened by exchange.
Heaters permitted by the Mayor,
Allowing Jack to shed a layer.
Dinners fit for an emperor,
With sushi and with tempura —
And hosts who are both warm of heart
And brilliant at the research art.

Thank you for everything.
Paul D. Bartlett

Paul D. Bartlett's impressions of Japan and the Japan–U.S. Seminar were written in rhyme. Because of the need to conserve fuel, the heat was turned off each year on April 1 in all national institutions in southwestern Japan. Professor John Roberts, the "Jack" referred to in line 16, was feeling the spring chill and had returned to his hotel to pick up a warm coat. Unbeknownst to him the conference planners had requested permission from the city authorities to install a heater in the conference hall. Permission was granted and the heater was installed while Dr. Roberts was fetching his coat. Upon his return, he found the hall warm enough "to shed a layer".

Riko Majima on his 88th birthday at his residence in Jurakuso, Takarazuka, in 1961.

A. 米 B. 米 C. 八十八 D. 八

Several birthdays are celebrated in Japan as special occasions: the 60th, 77th, 88th, 99th, and 100th. The symbolism in the Chinese characters used to denote the number 88 are what make that year special. A. The Chinese character for rice, which was intimately involved in every aspect of daily prewar Japanese life. B. A slight modification to this character changes it to a character that can be broken down into (C) the three characters that represent the numbers eighty eight. They are, from left to right, eight, ten, *and* eight. *Furthermore, the character for* eight *consists of (D) two brush strokes that widen at one end, implying a broadened or prosperous future. Thus, 88 in Chinese characters not only stands for rice but also for a prosperous future (in duplicate).*

Celebrating the 30th anniversary of the journal Complete Chemical Abstracts of Japan *in Osaka, November 1956. First row from left: Professor I. Nitta (physical chemistry, Osaka University), Professor M. Kotake (organic chemistry, Osaka University), Dr. M. Sato (former president of the Central Research Institute of the Manchurian Railway), Professor Riko Majima (president of Osaka University), Dr. H. Takaoka (president of Toyokoatsu & Co.), Professor R. Nozu (organic chemistry, Kyoto University), and Professor S. Fujise (organic chemistry, Tohoku University). Second row from left: 2nd, Professor S. Akabori (organic chemistry, Osaka University), Professor T. Watase (analytical chemistry, Osaka University), Professor S. Kubota (organic chemistry, Osaka City University), T. Kaneko (biochemistry, Osaka University), and Professor T. Sakan (organic chemistry, Osaka City University). Takaoka, Kotake, Akabori, Kubota, and Kaneko were students of Professor Majima. (Photo courtesy of S. Akabori.)*

Majima Family Tree in Chemistry

Riko Majima (Tohoku University and Osaka University)

Professor Majima (1874–1962) graduated from Tokyo University Science University in July 1898 and subsequently earned the D.Sc. degree for his contributions to research. He investigated the ozonolysis of urushiol and other subjects from 1907 to 1910 under C. D. Harries at Kiel University in Germany, then spent another year and a half with R. Willstatter in Zurich and at the Davy Faraday Institute in London. He returned to Japan after 4 years and was appointed professor in 1911 at the new Tohoku Imperial University.

First Generation	*Second Generation*
Chika Kuroda (1915) (Tokyo Higher Normal School)	
Harusada Suginome (1918) (Hokkaido University, Sapporo)	Sumio Umezawa (Keio University, Tokyo) Toshi Irie (Hokkaido University, Sapporo) Takashi Matsumoto (Hokkaido University, Sapporo) Tadashi Masamune (Hokkaido University, Sapporo) Hiroshi Suginome (Hokkaido University, Sapporo) Masaji Ohno (University of Tokyo) Akira Suzuki (Hokkaido University, Sapporo)
Munio Kotake (1919) (Osaka University)	Takashi Kubota (Osaka City University) Yasuhide Yukawa (Osaka University) Takeo Sakan (Osaka City University) Masazumi Nakagawa (Osaka University) Masao Nakazaki (Osaka University) Tetsuo Shiba (Osaka University) Toshio Miwa (Osaka City University) Teruo Matsuura (Kyoto University) Soichi Misumi (Osaka University) Keizo Naya (Kanseigakuin University)
Shinichi Morio (1919) (Tohoku University)	
Shinichi Kawai (1922) (Tokyo University of Literature & Science)	Noboru Sugiyama (Tokyo University of Education) Masahiro Hatano (Tohoku University)
Terutaro Ogata (1922) (Institute of Physical & Chemical Research)	
Shinichiro Fujise (1923) (Tohoku University)	Sekio Mitsui (Tohoku University) Shinichi Sasaki (Toyohashi Technical College) Hisashi Uda (Tohoku University)

Continued on next page

First Generation	*Second Generation*
Toshio Hoshino (1924) (Tokyo Institute of Technology)	Yoshio Iwakura (Tokyo Institute of Technology)
	Masaki Ohta (Tokyo Institute of Technology)
	Kenzo Kurihara (Hokkaido University, Sapporo)
	Tetsuo Sato (Tokyo Institute of Technology)
	Teruaki Mukaiyama (University of Tokyo)
	Hisatsugu Yoshimura (Tokyo Institute of Technology)
	Oyo Mitsunobu (Aoyamagakuin University, Tokyo)
Shiro Akabori (1925) (Osaka University)	Yoshio Matsushima (Osaka University)
	Nobue Hagihara (Osaka University)
	Seinosuke Otsuka (Osaka University)
	Yuichi Yamamura (Osaka University)
	Ryo Sato (Osaka University)
	Yoshiharu Izumi (Osaka University)
	Nobuo Tamiya (Tohoku University)
	Kozo Narita (Osaka University)
	Hidesaburo Hanabusa (Rockefeller University)
	Masatsugu Inoue (New York State University)
	Yoshimi Okada (Tokyo University)
	Tokuji Ikenaka (Osaka University)
	Masayasu Nomura (University of Wisconsin)
	Hiroshi Kotake (Kanazawa University)
	Horishi Tsugita (Tokyo Science University)
	Fumio Sakiyama (Osaka University)
Masuo Murakami (1926) (Osaka University)	Yasuhide Yukawa (Osaka University)
	Shigeru Oae (Tsukuba University)
	Ichiro Moritani (Osaka University)
	Yoshiyuki Okamoto (New York Institute of Technology)
	Shinya Nishida (Hokkaido University)
Tetsuo Nozoe (1926) (Tohoku University)	Akira Yokoo (Okayama University)
	Shuichi Seto (Tohoku University)
	Yoshio Kitahara (Tohoku University)
	Toshio Mukai (Tohoku University)
	Shô Itô (Tohoku University)
	Akira Yoshikoshi (Tohoku University)
	Kahei Takase (Tohoku University)
	Satoru Masamune (Massachusetts Institute of Technology)
	Ichiro Murata (Osaka University)
	Yusaku Ikegami (Tohoku University)
	Toyonobu Asao (Tohoku University)
	Hitoshi Takeshita (Kyushu University)
	Shigeo Nozoe (Tohoku University)
	Kyozo Ogura (Tohoku University)
	Tsutomu Miyashi (Tohoku University)
	Masaji Oda (Osaka University)

Continued on next page

First Generation	*Second Generation*
Eiichi Funakubo (1928) (Osaka University)	Minoru Imoto (Osaka City University)
	Ichiro Moritani (Osaka University)
	Niichiro Tokura (Tohoku University)
	Eiji Imoto (Osaka Prefecture University)
	Horoshi Taniguchi (Kyushu University)
	Shun-ichi Murahashi (Osaka University)
Shunsuke Murahashi (1930) (Osaka University)	Nobue Hagihara (Osaka University)
	Seinosuke Otsuka (Osaka University)
	Heimei Yuki (Osaka University)
	Hiroshi Yamazaki (Institute of Physical & Chemical Research)
	Akira Nakamura (Osaka University)
Kunisaburo Tamura (1930) (Institute of Physical & Chemical Research)	
Takeo Kaneko (1931) (Osaka University)	Masazumi Nakagawa (Osaka University)
	Tetsuo Shiba (Osaka University)
	Kaoru Harada (Tsukuba University)
	Terukiyo Hanafusa (Osaka University)
	Takashi Amiya (Hokkaido University, Sapporo)
	Shoichi Kusumoto (Osaka University)
Takashi Kubota (1932)	Keizo Naya (Kanseigakuin University)
	Teruo Matsuura (Kyoto University)
	Isao Kubo (University of California—Berkeley)
	Yoko Naya (Suntory Institute of Bioorganic Sciences)
	Takashi Tokoroyama (Osaka City University)
	Tadao Kamiyama (Kinki University)

NOTES: I constructed the Majima tree following an earnest request from the editor. It is quite possible that the names of many important chemists are missing. I take full responsibility for such inadvertent omissions.

In prewar Japan, it was customary to submit the Ph.D. thesis after many years of research and after being appointed to the rank of assistant and associate professor; thus, the years listed in this tree generally denote the year of graduation with a B.Sc.

This tree lists people in academic positions only and does not include scientists who joined industry.

and publish reports of our work in English. The papers on tropolone chemistry that were kindly communicated almost every month by Professor Majima to the Japan Academy began to appear in July 1950.

The reports were translated by Miss Dorothy U. Mizoguchi, because I was not experienced in writing in English. Dorothy Mizoguchi was a daughter of a viscount. She had been in London with her

Dorothy U. Mizoguchi's visit to the chemistry department of Tohoku University. Photo taken at Gonryo Kaikan in Sendai, 1958. First row from left: T. Mukai, D. U. Mizoguchi, Takako Nozoe, Y. Mayama, T. Sato, and K. Takase. Second row from left: Yoshio Kitahara, T. Ikemi, Tetsuo Nozoe, Shuichi Seto, I. Murata, S. Matsumura, and Shô Itô.

parents during her middle-school years. After her parents returned to Tokyo, she graduated from Tokyo Women's College of Pharmacy. Because of her unique career, she wrote papers on chemistry and published an introductory book for chemists for writing papers in English. She was a member of the group under Professor Eiji Ochiai at the Department of Pharmacy in Tokyo University and helped translate publications for the Japan Pharmaceutical Society. I was introduced to her by Professor Ochiai and I asked her to help my group translate papers and lecture drafts. I will not forget the help extended by Miss Mizoguchi, who carried out an invaluable translation of my earlier works in English. Unfortunately, she died abroad in a drowning accident.

Because the *Proceedings of the Japan Academy* had a very limited circulation, I took the initiative of sending our reprints to fellow researchers abroad who were conducting the same research as we were and also to some of the key figures of organic chemistry in foreign countries. This initiative led to a brisk and welcome exchange of publications between ourselves and the overseas chemists.

Unexpectedly, in the summer of 1952, I received an invitation from Professor Erdtman of Sweden to give a lecture on hinokitiol at the Symposium of Natural Products Chemistry. This symposium was to be a part of the IUPAC Congress that was being held in Stockholm in the

On the occasion of Professor and Mrs. Erdtman's first visit to Japan in June 1958, Dr. Katsura and Nozoe took them to the Hiba Forest in Aomori prefecture. Professor Holger Erdtman (kneeling) investigates a grub containing hinokitiol. Behind him from left: Tetsuo Nozoe, Mrs. Erdtman, Shigeo Katsura, and a guide from the Akita Timber Company.

summer of the following year. I gladly accepted the invitation and applied to the Japan Science Council for permission to attend the symposium. Fortunately, the council granted my request to be sent as one of the delegates from Japan. However, at that time, foreign currency (U.S. dollars) was hard to obtain, and a delegate who was attending an international conference was given an allowance only for a round-trip air fare and living expenses during the conference. The allowance for foreign currency was thus confined to the attendance of conferences; no other official or private trips were applicable. I negotiated with the Ministry of Education so that I could obtain a sufficient amount of foreign currency for private trips to talk directly with researchers in both Europe and North America. Finally, just 1 month prior to my scheduled departure, I was allowed to exchange my Japanese money into U.S. dollars sufficient for a 4-month private trip, which would allow me to visit many researchers abroad.

Because this overseas trip was to be a new experience for me, I decided to contact those professors whom I knew through the exchange of publications and ask them to assist me in setting up an itinerary. I wrote to Professor Clemens Schöpf (Darmstadt) regarding a trip to Germany, Professor Alexander Todd (Cambridge) for the United Kingdom, Professor Du Pont (École Normale Supérieur) for France, and Professor Louis Fieser (Harvard) for the United States. In each letter, I indicated the flight number, the airport at which I would arrive, the number of days I was planning to spend, and the names of the persons with whom I wished to meet. Within approximately 2 weeks, every one of my letters had been kindly answered. Each person indicated that because it would take some time for a detailed schedule to be drawn up, they would contact me at the respective airports or through the IUPAC Congress Office in Stockholm. I was also asked to prepare a 1-hour lecture based on our research in tropolone chemistry, in addition to the lecture for the symposium in Stockholm.

Again, Dorothy Mizoguchi translated the manuscript and Dr. Koji Nakanishi recorded it on tape. I spent the remainder of my time before the trip practicing my English pronunciation and intonation via the tape and Linguaphone record, while Toshiaki Ikemi prepared very beautiful, hand-written slides for my presentations.

I left Japan being certain only of my arrival time in Frankfurt and the schedule of events at the congress in Stockholm. To make things even worse, my flight was postponed unexpectedly because of a storm. I asked the airline to send a telegram to Professor Schöpf to inform him of my 1-day delay. I had originally planned to arrive in Europe on a Friday, because I had been advised that Saturdays and Sundays were not suitable for shopping or visiting the laboratories. However, with the delay, I ended up arriving in Frankfurt on the evening of Saturday,

This photo was taken in 1953 at the 14th IUPAC Congress banquet in Stockholm during Nozoe's first overseas tour. Torsyen Althin (center) was the director of the Swedish Technical Museum; he introduced Nozoe to Mr. Persson, the so-called "Persson in Skabersjö" and Minister of Education. (Identified by Mrs. Aulin-Erdtman and Professor Wachmeister, University of Stockholm.)

July 20. Luck was with me, because I found that Professor Schöpf had, indeed, been contacted at the airport and had left a message telling me to phone him at his residence as soon as I arrived. We made arrangements to meet the next Sunday morning, and I was taken on a tour of the university's empty laboratories. I was then received warmly at a luncheon by Professor Schöpf and his wife at their residence.

Before I left, I handed a notebook to Professor Schöpf and asked him to scribble something in it in memory of our meeting. He wrote, *"Ein liebes Besuch aus einem schönen, fremden Land erfreute uns und gezeigt wie völkerverbindet die Wissenschaft ist. Viele gute Wünsche für die weitere Reise!"* with his and his wife's signatures (p 114). Those words were the beginning of a souvenir that would become an important memoir of my travels. Samples were selected for this book from an enormous number of signatures in my autograph book.

Following my arrival in Frankfurt, everything was taken care of by the professors and staffs of each university. I did not need to worry about hotel reservations, fixing appointments, or even transportation to and from the universities. Things were definitely not running as I had anticipated—they were going much too smoothly.

My next stop was Heidelberg, where I was met by Professor O. T. Schmidt. Professor Schmidt took me on a tour of Heidelberg

Ein lieber Besuch aus einem schönen, fremden Land erfreute uns und zeigt wie völkerverbindend die Wissenschaft ist. Viele gute Wünsche für die weitere Reise!

Darmstadt, den 19. 7. 1953. Charlotte Schöpf. C. Schöpf

Heidelberg.

Karl Freudenberg O. Th. Schmidt.

Einen herzlichen Gruss dem japanischen Kollegen und gleichzeitig dem grossen Lande Japan.

Tübingen, 21. Juli 1953 Georg Wittig Eug. Müller

F. Weygand Rudolf Grewe

Vielen Dank für den freundlichen Besuch und die besten Wünsche für die Diskussionen in Stockholm

Tübingen, 21. Juli 1953 Walter Hückel

Wir hatten heute die grosse Vergnügen, Herrn Prof. Dr. Nozoe, Sendai aus dem Sonnenland Japan bei uns begrüßen zu können. [illegible] Räume [illegible] Prof. Nozoe [illegible] Arbeitsgebiet!

Ludwigshafen a/Rh., d. 28. 8. 53. Dr. Walter Reppe

A remembrance of the author's first visit to Germany in July 1953. The comments written by Professor and Mrs. Schöpf of Darmstadt are followed by autographs of many professors from various places: Heidelberg (K. Freudenberg and O. Th. Schmidt); Tübingen (F. Weygand, E. Müller, G. Wittig, and W. Hückel); Kiel (R. Grewe); and BASF (W. Reppe).

University laboratories, where I met both Professors K. Freudenberg and H. Lettrè of the Cancer Research Institute. The following day, I was guided by Professor F. Weygand through Tübingen University. The lecture in this university marked my first lecture given in English. In the new, magnificent lecture hall were seated more than 200 students and staff members, including Professors Eugen Müller, Georg Wittig, and Walter Hückel in the front row. With my manuscript for the lecture, entitled "Recent Advances in Troponoid Chemistry," in hand, I stepped onto the platform.

When my slides were projected on a large white wall as screen, I was momentarily impressed by their effect. Such a sophisticated projector or screen was not available in Japan at that time. Following a cordial introduction by Professor Weygand, I summoned my courage and began my lecture with a clear voice. My concluding statement, "I assume that some day tropolone chemistry will contribute not only to organic chemistry but also to the welfare of mankind," brought a chorus of applause. Professor Weygand later told me that the audience had been greatly impressed. I was happy enough that my accented English had been well understood.

My second lecture in Germany was given at the University of Kiel. This was the same institute in which Professor Majima had once studied the ozonization reactions of urushiol with Professor C. D. Harries. I was one of two lecturers invited to this seminar by the northern branch of the German Chemical Society. After his presentation on protein, the first lecturer, Dr. H. Zahn was presented with a gift, a string of sausages tied with a red ribbon. After my talk on tropolone, the chairman, Professor Rudolf Grewe, thanked me for my speech and presented me with a small white box tied with a red ribbon. With all eyes fixed upon me, I opened the box and found a small silver heptagon-shaped ashtray with the inscription, "Prof. T. Nozoe, *Zur Erinnerung an Kiel,* 24. 7. 1953." I was told that the ashtray was made of copper gilded with nickel (prepared at the university's mechanics shop), and its shape was a replica of that of tropolone with the conjugated double bond.

I was impressed with both the sense of humor and the hospitality of these people. Their way of running a seminar was certainly different from the Japanese way. In those days in Japan, almost no questions were asked by the audience after the lectures, and only short, polite acknowledgments were given by the chairman. Today, however, as many young people have had more opportunities to visit or study in the United States and Europe, questions and discussions after lectures have become common. Nevertheless, it seems to me that such a custom is still less intense in Japan than it is abroad, most likely owing to the language barrier in English and our commonly observed characteristic hesitation in clearly expressing our opinions in public.

On the 26th of July, I finally arrived in Stockholm. The next morning, as I made my way to the IUPAC Congress, I was surprised to meet, one after another, my very limited friends and many of the people I had hoped to meet. Sitting behind me on the bus were Professor Schöpf and his wife. As I stepped off the bus, Professor Roger Adams stood close by. When I went over to the registration desk to register, I was stopped by an unfamiliar person who addressed me by name and asked if I would give a lecture at his institute. When I asked which university that might be, he answered, "ETH-Zurich," and introduced himself as Professor Ruzicka, "Ruzicka, pronounced just like rouge for ladies," he said, pointing at his lips. Then because I had wanted to visit Professor Erdtman, I had asked another gentleman at the campus where I could find Professor Erdtman's office. Without answering my question, he asked, "How is your research on hinokitiol going?" He was Professor Erdtman himself. I began to realize exactly how small the society of scientific researchers really was! The warm hospitality, together with the wonderful sense of humor of the people I met, helped to alleviate the weariness and tension associated with my first travel abroad.

The first plenary lecture at the conference was given by Professor Linus Pauling, who won the Nobel Prize in chemistry the next year. Professor Pauling's presentation fascinated the audience and filled me with admiration. As I have stated before, my tropolone research in Formosa was heavily indebted to his book.[36] After this lecture, at a dinner hosted by Professor and Mrs. Erdtman, I had the chance to converse with Professors P. A. Plattner and V. Prelog (Zurich), H. Lettrè (Heidelberg), F. Šorm (Prague), and A. G. Anderson, Jr. (San Francisco) about work related to my research. During my stay in Sweden, I also had the chance to converse with Nobel laureates, Professor A. Tiselius (President of the IUPAC Congress), T. Svedberg, O. Virtanen, H. Theorell, Otto Hahn, and Hans von Euler (at his residence).

After the conference, I first traveled to several countries in northern Europe (Norway, Denmark, and Finland) for sightseeing during the summer vacation. Then I began my scientific visit to various countries in Europe and the United States. In Germany, I met Dr. W. Reppe in his office at BASF (Badische Anilin- und Soda Fabrik, AG), in Ludwigshaven. I was astonished to find that a large carpet in his office was decorated with the eight-membered structure of cyclooctatetraene.

Upon traveling to Switzerland, I visited the ETH in Zurich and the University of Zurich, as well as four pharmaceutical companies in Basel, in 3 days. I met Professors P. Karrer, T. Reichstein, C. Neuberg, Ruzicka, Prelog, Plattner, and A. Eschenmoser. Even at the ETH, our

Arne Tiselius
Börje Wickberg

Gunhild Aulin-Erdtman 27.7.53

Med önskan om fortsatt framgång med tropolonkemin
With wishes of continous progress in tropolon chemistry
Bengt Lindberg

Aug. 19. Hugo Theorell
Stockholm, Nobel Institute of Biochemistry
Aug. 19, [illegible], Stockholm
Nobel Institute of Biochemistry.

Hope to see you again
Uppsala, August 8th
Arne Fredga

Erik Jorpes
Karolinska institutet
Stockholm.

Einar Stenhagen

Shinzo Koizumi
小泉信三 小泉とみ

20. VIII 1953 Hans von Euler
University of Stockholm.
Institute for Research in organic Chemistry

Nozoe's first trip to Sweden, July 1953. He visited the Royal Institute of Technology (Professor and Mrs. Erdtman and Dr. Wickberg); University of Stockholm (Professors Lindberg and von Euler); Nobel Institute of Biochemistry (Professors Jorpes and Theorell); Uppsala (Professors Tiselius, Fredga, and Stenhagen); and Dr. Shinzo Koizumi (tutor of Crown Prince Akihito, now emperor of Japan) and Mrs. Koizumi.

topics of discussions were restricted to tropolones and azulenes (not sapogenins).

In France, I was met by Professor Du Pont. Accompanying him was his young assistant, Dr. Guy Ourisson. While in Paris, I visited the Pasteur Institute, the College de France, and the French Academy.

NaCl → DBV⁻Na⁺ ← CaCl$_2$

CaCl$_2$ ← DBV⁻ Ca⁺Cl⁻ → NaCl

Tokyo, March 30. 1978. Stereospecific selections from [illegible] of Organic Chemistry and Playboy (72)

Professor Vladimir Prelog was decorated with the Second Order of the Sacred Treasure in Tokyo, April 1977. Joining him at a celebration party, from left: Professor Hamao Umezawa, K. Tsuda, Vlado Prelog, Professor Teiji Ukai, Mrs. Prelog, and Tetsuo Nozoe. This autograph was obtained during Prelog's visit to Japan to attend the Centennial of the Chemical Society of Japan as an honorary member in March 1978. Because he gave no lecture on that occasion, he signed as "Playboy (72-year-old)".

Zur freundlichen Erinnerung an Ihren liebenswürdigen Besuch in unserem Laboratorium und an Ihren schönen Vortrag über Tropolone in unserem Kolloquium!

Zürich, 11.IX.1953 L. Ruzicka

We all like Professor Nozoe and his tropolones

V. Prelog

E. Heilbronner A. Eschenmoser

Pro memoria Tropolones Arthur Stoll

Basel, den 15-9-53 T. Reichstein C. Grob.

H. Schmid

Pl. A. Plattner
14. Sept. 1953.

Ich danke Ihnen für Ihren freundlichen Besuch und werde mich stets gerne an ihn erinnern.

Zürich, 16. September 1953. P. Karrer.

Nozoe was introduced to Switzerland in September 1953: ETH Zürich (Professors Ruzicka, Prelog, Heilbronner, and Eschenmoser); University of Zürich (Professors Karrer and Schmidt); University of Basel (Reichstein and Grob); and private industry (Drs. Plattner and A. Stoll).

Très heureux et honoré de recevoir dans notre laboratoire de Chimie de l'École Normale Supérieure de Paris le Professeur Nozoe — je souhaite que cette visite soit l'origine d'une collaboration internationale entre les deux laboratoires

Guy Ourisson

Nous nous souviendrons longtemps avec plaisir de la visite du Professeur Nozoé à l'Institut du Radium de l'Université de Paris, et lui exprimons nos meilleurs voeux pour la brillante continuation de ses remarquables travaux.

Zajdela (Biologiste à l'Institut du Radium) E. Lederer N.P. Buu-Hoï 宝薈

Paris, le 17.9.1953.

Pascaline Daudel (Institut du Radium Paris) Rigaudy

Le rubrène est très flatté de la visite qu'est venu lui faire le Professeur Nozoe, moi aussi!

Dufraisse

France followed Switzerland on Nozoe's travel agenda. He toured the Ècole Normale Supérieure (Professor Dupont, president, and Dr. Ourisson); Radium Institute (Professor Buhoi); Institute of Biology and Physical Chemistry (Professor Lederer); and other institutes and academies (Professor Duffraise and Dr. Rigaudy).

After leaving Paris, I traveled to London. At the airport, I received a very busy schedule set up by Dr. Alan Johnson (Cambridge University), who was researching natural tropolone metabolites under Professor Alexander R. Todd. At that time, Professor Todd was in Moscow, and so on his behalf, Dr. Johnson made arrangements for me. On my first day in London, I visited the Royal Cancer Hospital, the Lister Institute, and Birkbeck College. At Birkbeck, I was asked by Professor D. H. R. Barton to give a lecture on tropolone chemistry. I was the first Japanese person he had ever met, and our friendship continues to this day. My second day in England was spent at Queen Mary College, where I discussed tropolone and hinokitiol with Professor M. J. S. Dewar. Then, I proceeded to the laboratory of Professor Raistrick, the discoverer of stipitatic acid. Afterwards, I visited both Oxford and Cambridge Universities. At Cambridge, I was asked by Dr. Johnson to repeat my tropolone lecture. Dr. Johnson's main interest of study also happened to be tropolone chemistry, so he was quite intrigued with the rare samples of troponoid derivatives and colorful hinopurpurins that I had brought with me in three small colorful cigarette cases. After the lecture, I visited Sheffield University, where I met with Professor R. D. Haworth and Dr. P. L. Pauson, who had just returned from the United States. On that day, Professor Haworth kindly invited me to stay at his residence. At that time, I did not realize that he was the very man who had written the aforementioned review on sapogenins in the *Annual Report*,[9] hence we talked only about tropolone on that occasion.

The final event of my British trip was the visit to Professor J. W. Cook at Glasgow University. It was there I had the opportunity to speak to many of the young researchers whose names I had only seen in references. One of the key staff members, Dr. R. Raphael, wrote in my autograph book, "After hearing Prof. Nozoe, I think tropolone must now bear the trademark, 'Made in Japan'." On the afternoon of the last Sunday in the United Kingdom, after visiting Professor Todd, who had just returned from Moscow, I left for the United States.

My schedule in the United States was much the same as that in Europe. It was very busy and included visits to pharmaceutical companies and laboratories related to cancer research and many receptions and invitations at professors' residences. One difference that I did notice, however, was the greater number of Japanese chemists that I met, compared with those in Europe. In all of the major universities and laboratories in the United States that I visited, there were Japanese students working toward postdoctoral or postgraduate degrees in chemistry. In Europe, I met only six Japanese, including the Japanese Crown Prince who was visiting Stockholm after the British coronation ceremony.

With very best thanks for a most interesting lecture and best wishes for your future work.

D. H. R. Barton.

Birkbeck College, University
Sept. 22nd, 1953, London.

It was exciting news to hear that Prof. Nozoe was visiting this country and I have looked forward enormously to his visit. I have enjoyed meeting you: and all good wishes for the future.

Michael J. S. Dewar

Queen Mary College, University of London, 23.ix.53

With many thanks for your visit & the interesting discussions. Best wishes for future success in the tropolone field

28th September 1953
Chemistry Dept.,
The University,
Sheffield 10

R. D. Haworth.

T. S. Stevens. P. L. Pauson

Dorothy Haworth

R. P. Linstead E. A. Braude

24. September 1953
Imperial College of Science and Technology,
London, SW7

Pages from Nozoe's first visit to the United Kingdom in September 1953. University of London (Professors Barton and Dewar); Imperial College (Professor Linstead and Dr. Braude); Sheffield (Professor Haworth, Dr. Stevens, and Dr. Pauson). Dr. Pauson, who had just come back from the United States, wrote a detailed review in 1955 of the author's work on tropolone chemistry.

Left: Lord Todd (Nobel Prize, 1957) signed Nozoe's book again in Tokyo on September 22, 1965. Bottom: More mementoes of the United Kingdom in 1953. Cambridge (Lord Todd's group, Drs. A. W. Johnson and Kenner); Glasgow (Professor Cook, Drs. Loudon and Raphael). On the last day of his trip Nozoe visited the residence of Lord Todd.

We greatly enjoyed the excellent lecture on Tropolones which Professor Nozoe has given in Glasgow University and we delight to make his acquaintance

J. W. Cook.
Geo. Buchanan
J. Loudon

After hearing Prof. Nozoe I think tropolone must now bear the trade-mark "Made in Japan".
(11/X/53)
R. Raphael

To Professor Nozoe who provided piquant sauce to flavour a British Railways sandwich.
Ex Converse Memorial Lab., Cambridge, Mass.
En route Chemical Lab., Cambridge, Eng.
V. M. Clark
25-IX-53

Professor Nozoe will teach us a lot about tropolones - and also how to carry our specimens when travelling.
G. W. Kenner.
A W Johnson.
University Chemical Laboratory, Cambridge

This unexpected but very great pleasure for us — a visit from Prof. Nozoe on a sunny Sunday afternoon in Cambridge. We hope our friendly contact thus made will continue & strengthen in the future!
4th October 1953.
32 Barrow Rd. Cambridge
Alex. R. Todd.
Alison Todd

My first U.S. stop was Cambridge, Massachusetts, where I attended a colloquium at the chemistry department of Harvard University. The chairman was Professor Fieser, who had arranged my extensive tour in the United States. Members of the audience present at the colloquium included Professors Bartlett, Bloch, Woodward, Westheimer, and Mary Fieser, as well as Professors Cope and Buchi of the Massachusetts Institute of Technology (MIT). From Harvard, I then proceeded to Yale University, where I met with Professor von Doering. Professor von Doering and his co-workers talked to me in detail about their research, because their work was very close to what we were doing in Japan. He also invited me to the well-equipped laboratories at Hickory Hill (Hichrill Chemical Research Foundation) to stay with them.

For the next 2 months, I traveled throughout a large part of the United States and Canada to visit the major universities and give lectures. I met many professors and was once again able to see my good friend Professor Roger Adams. I would say that I met almost all the major figures in organic chemistry in Europe and the United States. My last stop before my return to Japan was the University of Hawaii. On my return, I counted that I had visited more than 120 universities and laboratories and had given 26 lectures on troponoid chemistry. The signatures with structures and sentences in my autograph book then totaled 450; more than 10 of those people went on to become Nobel Prize winners in chemistry. I considered myself just a stranger from a faraway country, yet the people I met, the elite of the world of chemistry, were so very kind to me. The words Professor Schöpf wrote on the first page of my autograph book remain today very close to my heart.

Once I had settled back in Japan, visits by scientists from overseas (including L. Pauling, A. Butenandt, Derek Barton, Hans von Euler, R. Kuhn, and many others) became more frequent, and I received many requests for our reprints and to write reviews.

In 1957, I was again invited to the IUPAC Congress, this one to be held in Paris. This congress was to commemorate the centennial anniversary of the French Chemical Society. I was fortunate to be asked to be one of the main lecturers in the Organic Chemistry Division, along with Professors Marion (Ottawa), Prelog (Zurich), and Wittig (Heidelberg). At this conference, I met Professor Shemyakin (Moscow), who gave me a number of reprints on troponoid chemistry published in his country. My first trip overseas whetted my appetite for subsequent travels. I continued to visit foreign countries whenever I was given the chance. As a result, my autograph book, which was initiated by Professor Schöpf in 1953, is now the first of an eleven-volume set. As we chemists got to know each other better, the writing in my autograph book gradually began to vary, from structural formulas and chemical equations to poetic writings, photographs, or hand-drawn cartoons.

It is a pleasure to welcome you on your arrival and to wish you happy visiting in our country. American chemists already know your name and works, and will be all the more pleased to meet you personally.

Harvard Univ. Louis Fieser

October 8/53 Mary Fieser

RBWoodward

Arthur C. Cope George Büchi

May this be the first of many visits to this country. Paul D Bartlett

With best wishes to a truly great organic chemist.

Max Tishler William von Eggers Doering

Merck & Co Inc. Oct 10, 1953.

With great admiration and friendship

Oct. 24, 1953 Ray Pepinsky

State College, Pa. from me, too! Louise Pepinsky

To a distinguished worker in the tropolone field, with best regards –

D. S. Tarbell

University of Rochester

Oct. 26, 1953.

Souvenirs from Nozoe's first visit to the United States in October 1953. Harvard (Professors Fieser, Woodward, and Bartlett); MIT (Professors Cope and Büchi); Yale (Professor Doering); Pennsylvania State University (Professor Pepinsky); University of Rochester (Professor Tarbell); and Merck (Dr. Tishler).

Tetsuo Nozoe took these two photographs in October 1953 during his first trip to the United States. Above: Louis and Mary Fieser, who described Nozoe's hinokitiol work in the early 1950s in their famous textbook, Advanced Organic Chemistry. *Below: Professor R. B. Woodward, who often sat in his Harvard office until midnight discussing tropolone chemistry with Nozoe.*

The Second Order of the Sacred Treasure is conferred upon R. B. Woodward by the Japanese government, Tokyo 1970. With him at the ceremony are Tetsuo Nozoe and Sumio Umezawa, who is Hamao Umezawa's older brother.

At the celebration dinner in Tokyo for Professor G. Büchi when he became an honorary member of the Pharmaceutical Society of Japan in 1984. From left: Tetsuo Nozoe, Mrs. Büchi, Professor Satoru Masamune, Kyoko Nozoe, Mrs. Masamune, and the guest of honor.

A real honor to discuss tropolone chemistry with Prof. Nozoe. III

E. E. van Tamelen

Hyp J. Dauben Jr.
Dept. of Chemistry, Univ. of Wash.
Seattle

John D. Roberts

Donald J. Cram U.C.L.A. Chemistry Department, November 10, 1953.

A. G. Anderson, Jr.
Dept. of Chemistry
Univ. of Wash. Seattle

H. Rapoport
Berkeley, Calif.

It is a pleasure to meet again after our first acquaintance in Sendai

Roger Adams
Univ. of Illinois

William S. Johnson

Heartiest greetings from an inorganic chemist at Illinois
John C Bailar Jr.

Carl R. Noller
Stanford University

Jerome A. Berson

With a very warm Hawaiian aloha,

Leonora Neuffer Bilger, Chairman

November 24, 1953.

More souvenirs of Nozoe's first visit to the United States in October 1953. Many universities and companies were represented by the signatures of Professors H. J. Dauben, van Tamelen, Anderson, Roberts, Rapoport, R. Adams, Bailer, Cram, Noller, W. S. Johnson, Berson—and a "very warm Aloha" from Professor Bilger in Hawaii.

At a reception arranged by Nankodo Publishing Company to welcome Professor and Mrs. Linus Pauling, February 23, 1955, in Kamakura. Seated from left: Mr. M. Kodachi (president of Nankodo), Professor Linus Pauling (Nobel Prize, 1954), Dorothy U. Mizoguchi, Mrs. Pauling, Tetsuo Nozoe, and Mrs. Hayao. Standing from left: M. Komatsu (of Nankodo), Professor T. Shimanouchi, Professor S. Kikuchi, Dr. T. Noguchi, Dr. S. Hayao, and Professor Y. Miura at the far right.

Professor A. Butenandt (Nobel Prize, 1939), while visiting the department of chemistry at Tohoku University in Sendai, April 1955, signed the autograph book after seeing Nozoe's tropolone and azulene samples.

The 16th IUPAC Congress banquet at the Hall of Mirrors at Versailles, July 1957, on the occasion of the centennial of the Chemical Society of France. Sir Robert Robinson, as a representative of the participants, gave a speech. Clockwise from front center: 2nd, Professor L. Marion of Canada; 7th, Professor G. Wittig of Germany; 9th, Dr. R. Paul (scientific director of the Rhône Poulenc Company, later president of the Chemical Society of France); and 11th, Tetsuo Nozoe.

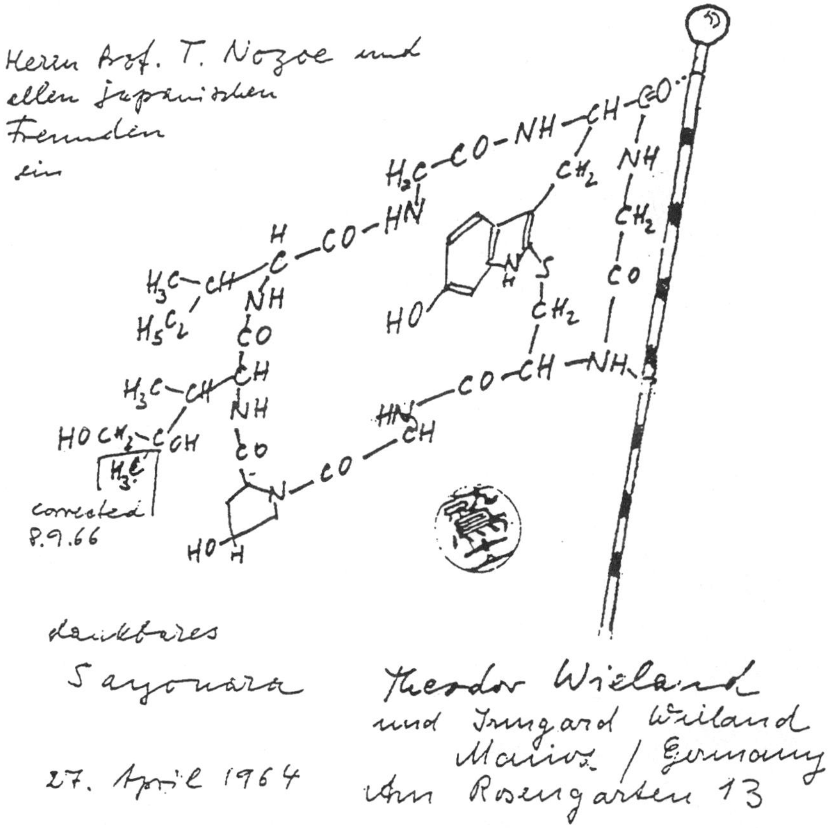

Professor T. Wieland visited Sendai in 1964. He wrote the structure of amanita toxin in the style of a Japanese flag, from which a methyl group was deleted during Nozoe's visit to him in Frankfurt in 1966. Wieland's Japanese seal was a present from Professor Munio Kotake.

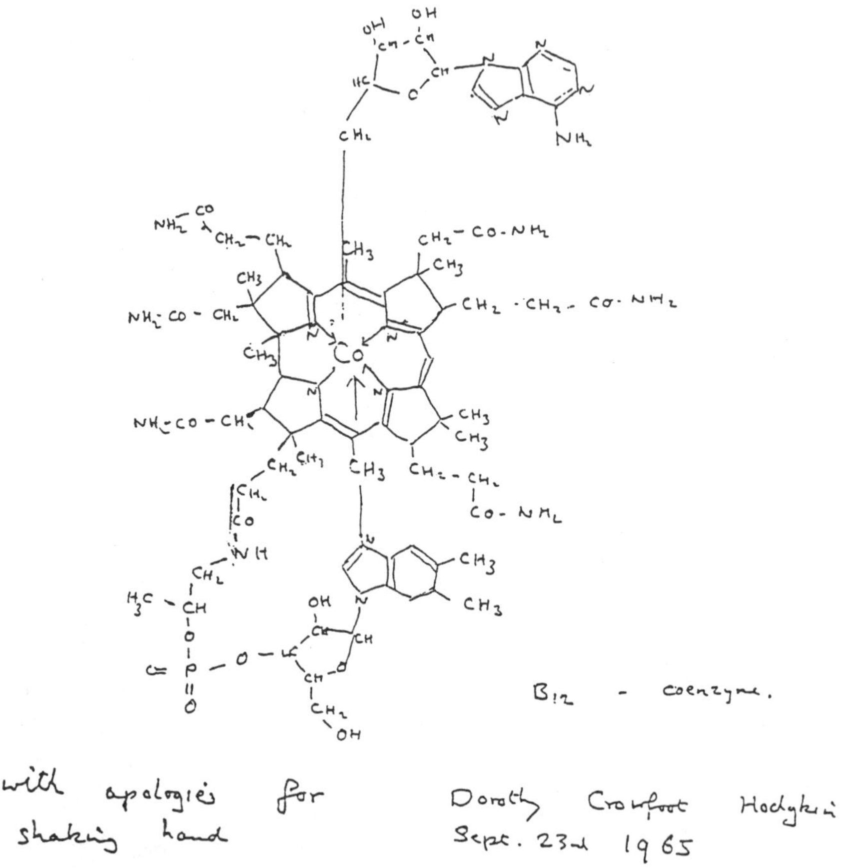

B_{12} - coenzyme.

with apologies for shaking hand

Dorothy Crowfoot Hodgkin
Sept. 23rd 1965

Dorothy Hodgkin visited Tokyo in 1965. She came to Japan, accompanied by a British princess and Lord Todd, as a cultural ambassador to the British exhibition held in Japan. Lady Hodgkin (Nobel Prize, 1964), an X-ray crystallographer, wrote for the first time this complex structure of B_{12}-coenzyme by hand. She spent a lot of time writing this, according to Miss Mizoguchi.

Molecular Recognition
→ Spherical Macrotricyclic Cryptand | A Tetrahedral Recognition Site

Jean-Marie Lehn —
April 20, 1979
U.L.P. Strasbourg

Nozoe visited Professor J.-M. Lehn at the University of Strasbourg in 1979. Lehn shared a Nobel Prize with D. J. Cram and C. J. Pedersen in 1987.

KON-TIKI

With cordial thanks for the wonderful [illegible] kind,
April 4. 1980

Professor Eschenmoser of ETH Zürich visited Tokyo in April 1980. He signed the autograph book after dinner with the Nozoes at Chinzanso.

It could not more enjoyable, as be together
with Mrs. and Mr. Nozoe ever so often!
Basel, 1. July 1985. R. Heilbronner

Professor Heilbronner of the University of Basel proposed a toast both to the completion of volume 8 of the author's autograph book, and to Nozoe and his wife, in Basel, 1985.

Although each of my trips seems to be on a tighter schedule than the last, I find the kindness shown to me, along with the enlightening research information, very refreshing. When asked if I am tired of traveling I always respond, "I have no time to be weary."

Honors and Awards. My first award, as I mentioned earlier, was the Majima Award during the war. The second award was the Asahi Cultural Award in 1952. The Asahi memorial lecture for my award was given by Professor Majima, who said, "Unsaturated seven-membered aromatic chemistry started because Tetsuo Nozoe (Tetsu-o means iron-man) discovered that the natural pigment hinokitin contains iron." The audience loved this remark, because it was an unusual joke from the usually very serious Professor Majima.

The following year, I was honored with the Japan Academy Award. After the ceremony and explanation to the Emperor of the recipients' achievements, we were invited to a luncheon in the palace. This amiable luncheon with the Emperor, Prince Mikasa, the Prime Minister, Shigeru Yoshida, the President of the Academy, Dr. S. Yamada, and the 10 recipients lasted for about 1 hour. This was one of the most impressive events in which I have participated. Around this time, the young crown prince was in Europe attending Queen Elizabeth's coronation. Because it was right after the war and there were anti-Japanese feelings among some of the British, Dr. Koizumi (a tutor to the crown prince) and his wife were suddenly dispatched to the United Kingdom. Prime Minister Yoshida had seen Dr. Koizumi off at Haneda Airport, and thus he was late for the luncheon. When he arrived he was seated

Tetsuo Nozoe's interview on the occasion of the Asahi Award in Sendai, January 1952. He was using the molecular model of hinokitiol to explain how the planar nucleus deviates to show olefinic characters in some cases.

Domo arrigato

Pine Island
23rd April
1964

coming to Japan
leaving again

Gunhild Aulin-Erdtman

A dinner for Professor and Mrs. Erdtman at Happoen, Tokyo, August 1958. Above, front row from left: Mrs. Gunhild Aulin-Erdtman, Professor Holger Erdtman, and Eiji Ochiai. Back row: Shoji Shibata, Tatsuo Kariyone, Tetsuo Nozoe, Shizuo Hattori, Kozo Miki, and Shigehiko Sugasawa. Below: Signature of Dr. and Mrs. Erdtman during their visit to Japan in 1964.

Receiving the Japan Academy Award in Tokyo in June 1953, preparatory to his presentation of his work to the emperor and guests at the Japan Academy. From left: Yoshio Kitahara, Tetsuo Nozoe, and Shuichi Seto.

next to the Emperor and across from me. The Emperor said to Yoshida, "Thank you," and then performed his duty as host to lighten the atmosphere by reminiscing about the time when he had attended King George VI's coronation and how warmly received and how happy he was when he first shopped by himself in England.

At this point, Yoshida, indicating to me, said to the Emperor, "This man is about to leave for an international meeting in Sweden." The Emperor then said, "It must be very pleasant to visit abroad and discuss science amongst your peers." President Yamada, despite being in the midst of a luncheon and, like most scholars, always thinking of nothing but research, said, "However, Your Highness, the government should pay for scientists to visit other countries in addition to the host country. Our government should strive to understand more about research." The Emperor then faced Yoshida and, in reply to President Yamada, said, "Discuss monetary matters with Yoshida." To this Yoshida jokingly replied, "I'm the Prime Minister, not the Minister of Finance." Everyone, including the Emperor, had a good laugh. Despite the austere atmosphere of the ceremony, the Emperor constantly brought up light topics of conversation, like the one just mentioned. After lunch, while drinking tea, all the recipients again explained their work to the Emperor, who asked questions of each of them, in turn. Of my research, he asked what the difference in constituents was between *taiwanhinoki* and the Japanese *hinoki*. As a biologist, he showed great interest in the fact that colchicine and hinokitiol had the same seven-membered skeleton.

In the fall of 1958, I received the most prestigious medal in Japan, the Order of Cultural Merit. I was not only overjoyed but also shocked. This award is the greatest honor in Japan and given only to four people every year. I first learned of this award as I was about to board a train at Tokyo Station to travel to the Second National Symposium on Natural Products Chemistry in Kyoto. Several press reporters were waiting for me at the train, because the announcement had been made after my departure from home. The two science recipients were Professor Heizaburo Kondo (pharmacy) and myself. It was the first time that both science winners were organic chemists. Up to that time, the only three recipients in chemistry were Professors Majima in 1949 (organic chemistry), Umetaro Suzuki in 1943 (agricultural chemistry), and Yasuhiko Asahina in 1943 (pharmaceutical chemistry). I was the first chemist to receive this honor before retirement. For some time, my surprise exceeded my joy, but then I became overwhelmed by the scientific community's understanding of my unique research area. In addition to rejoicing over my good fortune, I felt a strong responsibility as a scientist. Several years afterward, Professor Shiro Akabori, another student of Professor Majima, also received the Order of Cultural Merit. In 1959, I was presented with an honorary citizenship of Sendai.

Award of the Order of Cultural Merit from the Japanese government in Tokyo, November 3, 1958.

The person who recommended me to become a foreign member of the Royal Swedish Academy of Sciences (1972) was Professor Erdtman, who often jokingly referred to me as "my enemy". I was also recommended for an honorary membership of the Swiss Chemical Society by Professors E. Heilbronner, H. Zollinger, and V. Prelog, who were sometimes also engaged in the same area of research. Of the foreign awards I have received, the one that moved me most was the 1980 Wilhelm August von Hofman Memorial Medal, awarded in 1981. This medal was started in 1902 by Deutschen Chemischen Gesellschaft. The first three recipients were Professor H. Moisson, Sir William Ramsay, and Sir William Henry Perkin. I was the 30th recipient. Both the German professors Emanuel Vogel and Klaus Hafner, who were studying the similar field of nonbenzenoid aromatic chemistry, heartily congratulated me for this honor. Because I wished to present the first part of my award address in Hamburg in German, Professor and Mrs. Emanuel Vogel not only wrote my manuscript but also invited me to their room in the hotel and let me practice long into the night. My wife and I had been family friends of the Vogels and the Hafners for many years, and to this day, these friendships continue. I strongly agree with the inscription Professor Clemens Schöpf wrote in my autograph book in 1953 that science binds foreigners together (p 113).

Joint Research After Retirement from Tohoku University (1966–)

Development of Nonbenzenoid Aromatic Chemistry

Former Students and Colleagues. Since I returned from Taiwan and joined the faculty at Tohoku University, the number of my co-workers increased steadily, especially after the Japanese universities switched to the new system similar to that in the United States. My research group now consisted of 20–30 graduates and undergraduates and several technicians. Instead of restricting the research area to the chemistry of troponoids and azulenoids, we started some collaborative studies in biology and medicine with scientists of several universities, including Tohoku, Tokyo, and Osaka Universities. These studies were assisted financially by companies such as Takasago, Kao, and Sankyo, the support of which have continued since. These companies did not require me to collaborate with them. Even Dr. Yasota Kawakami and Seiji Sanga (currently a professor emeritus at Kansai University) assisted our research privately at the most trying period after my repatriation from Taiwan. Studies of troponoid and azulenoid chemistry expanded in our group, but unfortunately, investigations outside Japan were not so active as in the 1950s, except for those related to the cycloaddition and total synthesis of colchicine.

To extend further our studies in troponoid chemistry, I negotiated with the Ministry of Education for assistance. The support of this ministry led to the inauguration of two groups in troponoid chemistry within the Chemical Research Institute of Non-Aqueous Solutions, Tohoku University, and the appointments of Shuichi Seto and Yoshio Kitahara to the rank of full professor. Furthermore, arguing that Japanese educational institutions must be prepared for the development of future fundamental chemistry in Japan, I urged the Ministry of Education to expand our chemistry department. After several years of negotiation, this expansion was, fortunately, approved and the second department of chemistry consisting of six groups (*kôza*) started. Pro-

Tetsuo Nozoe and Dr. Hiraizumi at the 20th Symposium on the Chemistry of the Terpenes, Essential Oils, and Aromatics (TEAC) in Yamagata in October 1975. (Photo courtesy of H. Hirose.)

Discussion regarding troponoid of antitumor activity at the Research Institute in the Sankyo Company, Tokyo 1964. From left: Dr. H. Sato of Tohoku University, Tetsuo Nozoe, Dr. T. Matsui (an executive director at Sankyo Company), and Dr. Stock of the Sloan Kettering Institute in New York.

fessor Kitahara moved to the new department, whereas Toshio Mukai and Shô Itô were appointed professorships. Upon my retirement, Kahei Takase succeeded me in my group. I had to spend an enormous amount of time in connection with the numerous negotiations with the Ministry of Education in preparations for the new chemistry department. Consequently, at the time of my retirement, I was not prepared to continue my research.

After being promoted to full professor, Professor Seto began studying biochemistry. Besides the fundamental problems in tropolone chemistry, he also investigated the biosynthesis of the troponoid metabolite sepedonin, started biomedical studies, and initiated novel biosynthetic investigations of terpenoids with his colleague, Kyozo Ogura. My other colleague, Professor Kitahara, became actively engaged in studies of nonbenzenoid aromatic compounds and physiologically active natural products. Following my advice, he also moved into areas where I was weak, namely physical organic chemistry. He was bright, highly motivated, and also had a passion for drinking (not water, of course). His untimely death at the age of 53, when he was just about to embark on an exciting professional career, was a great loss to the scientific community.

Satoru Masamune was a classmate of Murata, and he studied the basic reactions of troponoids and preliminary synthetic investigations of simple colchicine analogues with Kitahara. Intelligent and extremely hard-working, Masamune's interest was academic, but realizing that my group was already overflowing with senior people and that future chances for him were dim, he secured a Fulbright fellowship and went to the University of California at Berkeley. There he received a Ph.D. with Professor H. Rapoport. At around this time, he wrote me that he wanted to marry my daughter; until then, I had not known that Takako was his girlfriend. After a postdoctoral period with Professor van Tamelen in Wisconsin, he joined the Mellon Institute in Pittsburgh and published his well-known three solo *JACS* communications on the total synthesis of the diterpenoid alkaloids veatchine and garryine (1964). He was contacted by Professor R. U. Lemieux (Alberta) during the 1964 IUPAC Natural Products Symposium held in Kyoto, and he moved to Canada (University of Alberta). Fourteen years later, he joined the faculty at MIT (Cambridge, MA). He tackles challenging projects. In the area of nonbenzenoid chemistry, he succeeded in the preparations of cyclobutadiene and [10]annulene. His main interest lies in the stereospecific synthesis of complex natural products and synthetic methodology. The future of chemistry is always an important factor in his research. His concepts are more American than Japanese.

My son Shigeo returned to Japan immediately after the war at the age of 14 on his own wishes, to which I consented because of the

Professor Mukaiyama's dinner in honor of Satoru Masamune's visit to Japan, December 2, 1980. Front row from left: Kyoko Nozoe, Satoru Masamune, and Tetsuo Nozoe. Back row from left: T. Mukaiyama, Mrs. Mukaiyama (Professor Hoshino's daughter), and Takako (Mrs. Masamune).

uncertainty of Taiwan's future at that time. However, not surprisingly, Shigeo was confronted with great troubles in postwar Japan. My brother's house in Osaka was sequestered by the American occupation force, and at one time, Shigeo had difficulty in finding a place to live. Moreover, he could not contact us in Taiwan by letters. Fortunately, Shigeo was finally able to find a space in Dr. Kafuku's temporary living quarters. Not only Dr. Kafuku's house in central Tokyo but also his place of refuge had been destroyed, and Dr. Kafuku and his family were living on a *tatami* (Japanese mat) in a corner of the laboratory in a suburb of Tokyo.

Shigeo joined us upon our return, and after graduating from a high school in Sendai, he went on to Tohoku University on his own initiative, studied troponoid chemistry with Professor Seto, and received his M.Sc. degree with me. Having determined to continue research on his own and not through the influence of his father, Shigeo hoped to be enrolled in a Ph.D. program at another university. Fortunately, Shigeo was able to join the group of Professor Kyosuke Tsuda, who had moved from Kyushu University to the Institute of Applied Microbiology,

University of Tokyo, where he finished his graduate studies. Shigeo went on to become an assistant and then an assistant professor. He spent a wonderful time at Harvard as a postdoctoral fellow with Professor E. J. Corey. Upon his return to Tokyo, Shigeo found that his institute had been especially hard hit for several years by the student riots that were rampaging throughout Japan. He lost precious time during a most important period of his career. Fortunately, in 1979 he became a professor in the Department of Pharmacy at Tohoku University.

Although they were not my students at Tohoku University, the two individuals deserve mention. Dr. Masato Tanabe is a Japanese-American and a treasured student of Professor W. G. Dauben (University of California at Berkeley). Tanabe spent half a year with me in 1955 on a Fulbright scholarship. Then, on my recommendation, he joined Professor Eiji Ochiai's group at the Department of Pharmacy, University of Tokyo, because I thought it would be beneficial for him to broaden his experience in another field. Although he addressed me as *Sensei* ("my professor" in Japanese), I treated him as an assistant professor in my group. I asked him to discuss research problems with members in my group and also to lecture on organic electronic theory, which was still quite new in Japan. He is a mild-natured and rather shy chemist, who always appeared to be enjoying life and chemistry. He currently

Farewell party for Dr. M. Tanabe at Shunpuso restaurant in Sendai, March 1955. Seated from left: Kyoko Nozoe, Kitahara's son, Tetsuo Nozoe, and Dr. Tanabe. Standing: Kitahara, Toda, Shigeo Nozoe, T. Ogata, Y. Mayama, and Takako Nozoe.

At a celebration in Tokyo for Professor William Dauben (University of California at Berkeley), appointed an honorary member of the Pharmaceutical Society of Japan, April 1987. From left: William and Carol Dauben, Tetsuo Nozoe, Mrs. Masamune, and Kyoko Nozoe.

occupies one of the top positions at the Stanford Research Institute and takes care of many Japanese chemists who visit that area.

Koji Nakanishi was a student of Professors Fujio Egami and Yoshimasa Hirata (Nagoya). Toshio Goto (Nagaya) and Yoshito Kishi (Harvard) are his juniors from the same school. I recommended Nakanishi as a successor to the chair of organic chemistry at Tohoku University upon retirement of my colleague, Professor Shinichiro Fujise. Nakanishi was at Harvard University with Professor L. Fieser as a GARIOR student for 2 years from 1950. He was one of the first chemists in postwar Japan to spend an extended time in the United States. His English is outstanding, and his personality is special. His main interests were in applying physical methods to structural determinations. He contributed greatly to our department by introducing new blood and equipment. He has accomplished outstanding research, exemplified by the finding and application of the intramolecular nuclear Overhauser effect during the structural studies of gingkolides. He has a unique social personality. In addition to his interests in chemistry, Nakanishi is also an accomplished performer of magic and delights us all during several social events.

After the 1964 IUPAC Natural Products Symposium in Sendai, I organized one of the postsymposium meetings. We held public lectures by three famous chemists (Professors G. Ourisson, F. Sondheimer, and E. Wenkert) in Sendai and followed this with the opening banquet and symposium at Sakunami, a spa on the outskirts of Sendai. In addition to the 150 graduate students and staff of organic chemistry of Tohoku University and other universities, company employees, as well as 60 overseas visitors and their families, attended the Sakunami meeting. Following the suggestion of Nakanishi, who was General Secretary, we all had Japanese-style dinner in light summer kimonos (*yukata*). Well-known chemists, including Professors D. Arigoni, D. H. R. Barton, D. Cram, C. Djerassi, H. Erdtman, N. Leonard, G. Stork, and their fam-

Before the Sakunami meeting, Sakunami station, May 1963. From left: Yoshio Kitahara, Tetsuo Nozoe, I. Murata, Shô Itô, and S. Sasaki. (Photo courtesy of K. Nakanishi.)

ilies were present. Nakanishi's hilarious after-dinner magic performance with his friend, Professor Ourisson, brought about a festive atmosphere, which led to many acts including group singing by the visitors' wives and daughters. Conversation between visiting scientists and our young graduate students continued into the early morning. The meeting was the envy of other chemists who later heard about it. ISNA-I (to be discussed later), held in 1970 in Sendai with me as Chairman and Shô Itô as General Secretary, was attended by many distinguished participants from overseas and was a great success. Although this meeting was not in his field, Nakanishi and his family came from New York at their own expense, and Koji was in charge of the dinner, night sessions, and other functions. His performance again created a lively atmosphere.

The opening dinner at the Sakunami meeting in Sendai, April 1964. Top, from left: Mrs. Gunhild Aulin-Erdtman, Tetsuo Nozoe, and Professor Holger Erdtman (Royal Institute of Technology in Stockholm). Bottom: Professor Leonard (University of Illinois), Tetsuo Nozoe, and Professor G. Ourisson (University of Strasbourg). Participants wore Japanese yukata *(kimonos).*

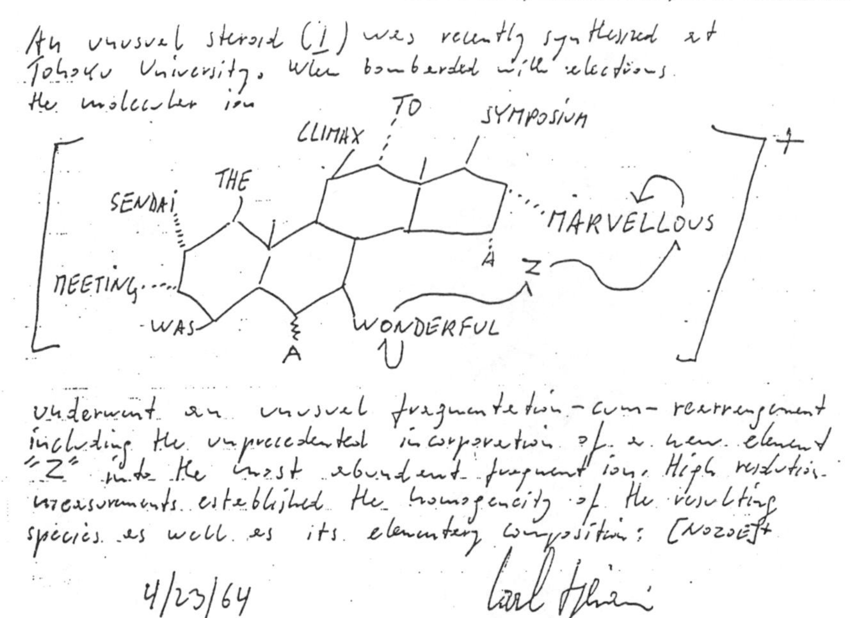
An unusual steroid (I) was recently synthesized at Tohoku University. When bombarded with electrons the molecular ion

underwent an unusual fragmentation-cum-rearrangement including the unprecedented incorporation of a new element "Z" into the most abundant fragment ion. High resolution measurements established the homogeneity of the resulting species as well as its elementary composition: [NOZOE]+

4/23/64 Carl Djerassi

Professor Carl Djerassi visited Sendai in April 1964 to attend a symposium in Sendai and Sakunami.

Around 1968, Professor Nakanishi received attractive offers from four universities in England, America, and Canada. Although usually not perturbed, he was at a loss this time, because besides the family factor, his work in Sendai, after 6 years, was starting to flourish. He visited my home in Tokyo several times to discuss these matters. Although I hated to see Sendai lose him, but taking his ability and personality in consideration, I finally agreed that it was in his best interests to go abroad, provided that a suitable successor at Sendai could be found. After repeated wavering, he finally decided to go to Columbia at the last moment. Professor Hideki Sakurai came to Sendai and has become a leading figure in organosilicon chemistry. The status of Japan was such that the language barrier and lack of social contacts of Japanese scientists made it difficult to carry out intimate collaborative studies with foreign scientists. However, Nakanishi was different, and I strongly believed that instead of being confined to the closed Japanese system, he should join a major university abroad and initiate original research. This would not only be beneficial for him but would also, in the long run, profit Japanese chemistry. When consulted, I agreed that he should leave. These expectations were fulfilled, and I am sure that if he had remained in Japan, his research would not have reached the same level.

At the welcome dinner for Professor R. Breslow, who visited Tohoku University in Sendai, April 1965. From left: T. Mukai, I. Murata, Yoshio Kitahara, R. Breslow, Koji Nakanishi, Tetsuo Nozoe, and Shuichi Seto.

The welcome dinner for Professor K. Bloch (Nobel Prize, 1964) at Tohoku University in Sendai, 1966. From left: Dr. B. J. Goffny (postdoctoral student in Nakanishi's laboratory), Professor Bloch, and Tetsuo Nozoe.

At the banquet of the 4th IUPAC International Conference on Organic Synthetic Chemistry in Tokyo, August 1982. From left: Professor Koji Nakanishi, Takako (Mrs. Masamune), Professor E. J. Corey, Tetsuo Nozoe, and Mrs. Nakanishi.

An alumni meeting of Taihoku Imperial University at Tatung Company in Taipei, December 1987. Dr. Nozoe visited Taiwan at the invitation of the National Research Council of Taiwan. Front row from left: Professors Y. T. Lin, L.-C. Lin, Z. S. Lin (vice-president), Dr. T. S. Lin (president), Tetsuo Nozoe, Kyoko Nozoe, Mrs. T. S. Lin, Professors S.-L. Liu, F.-C. Chen, and P.-W. Yang.

"Journal of Nozoe Organic Chemistry" (Journ. of premature results)
A journal with the "NO-REFEREE" system
Since 1953 → present, seven volumes published so far
Authors: Young, middle-aged, and old chemists
both in sober and drunken mental state
Subscription rate: nonexisting. (∵ no extra copies are availabl
Topics: Nonbenzenoid aromatics, and other subjects
both real ~~and~~ or irreproducible results
Chief Editor: Tetsuo Nozoe

" Sir: We hope to synthesize the following compounds 1 and 2

>N– ... CHO ... CHO

Although we have not yet succeeded in these syntheses, we are sending this to your Journal as we have heard rumors that all papers will be accepted. Hoping that you will treat this paper similarly,

Yours sincerely

9/20/1980

Koji Nakanishi

Professor Koji Nakanishi visited Nozoe's home in Tokyo in September 1980.

About 40 of the students I taught (including 10 in Taiwan) have become full professors. Many of them continued studying tropolones, azulenes, or their related compounds even after they had left my laboratory.

International Symposium on the Chemistry of Nonbenzenoid (Novel) Aromatic Compounds. In Japan and abroad, the study of troponoid chemistry made rapid progress, and by the late 1960s, the experimental verification of the Hückel rule became one of the central themes of physical organic chemistry. In addition to troponoids and azulenes, various annulenes, from three- or four-membered small rings to large 30-membered macrocyclic rings, were synthesized in rapid succession. Polycyclic and heterocyclic conjugated systems attracted much attention. Theoretical studies were also actively undertaken from the viewpoints of aromaticity or antiaromaticity. In this respect, it was helpful to be able to combine organic chemistry with quantum chemistry.

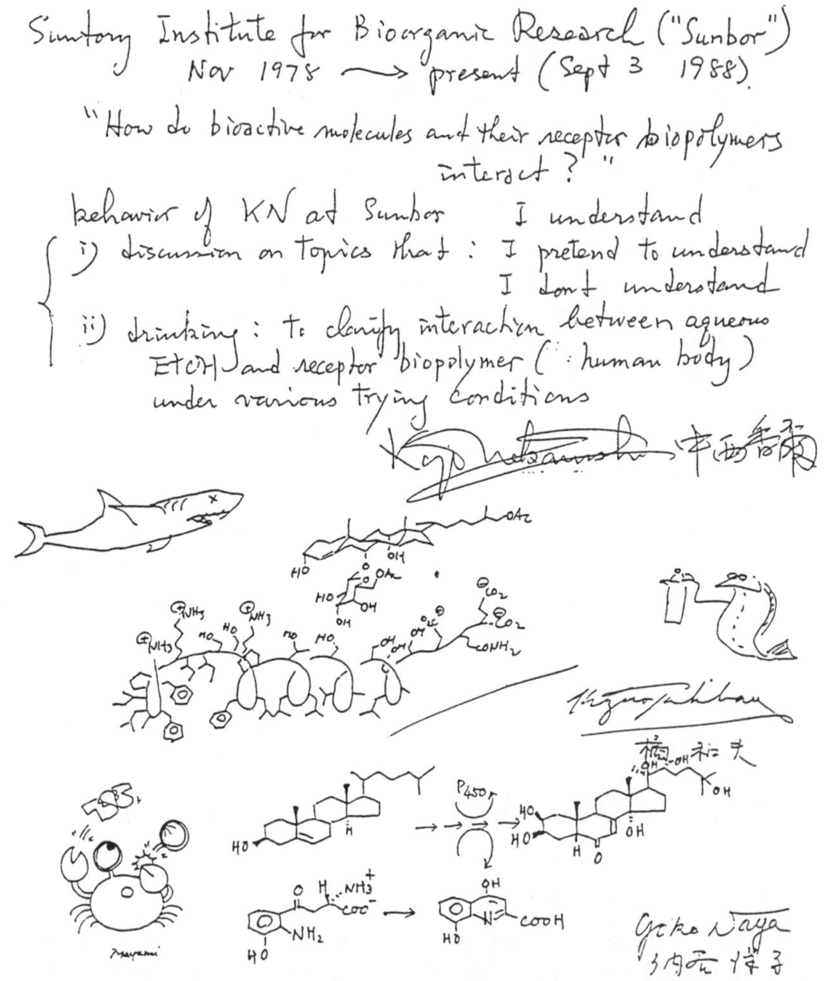

Suntory Institute of Bioorganic Research: Professor Koji Nakanishi (president), Dr. Yoko Naya (vice-president), and Dr. K. Tachibana (now a professor at Tokyo University), Shimamoto-cho, Osaka, September 1988.

Under these circumstances, the need for an international symposium became stronger, and finally, a suggestion of Professor R. Breslow led to the first IUPAC-sponsored International Symposium on the Chemistry of Nonbenzenoid Aromatic Compounds (ISNA). Organized by the Japan Science Council and the Chemical Society of Japan, this IUPAC symposium was held in Sendai in August 1970. I became the organizer of ISNA-1, and Shô Itô assisted me as the general secretary.

The ISNA was such a great success that it was decided to hold the symposium every 3– 4 years.

Naganin watakushi wa Nozoe kyoju to kyoju no kyodokenkyusha no minasama no kirei na oshigoto wo subarashii to omotte imashita. Gokofuru wo inorimasu!

Nozoe kyoju no to watakushi no kanshin wo kono kagobutsu ni mazatarerimasu!

ETH Zürich H. Zollinger

H. Zollinger, who had just been appointed an honorary professor at the Tokyo Institute of Technology, with the author in September 1983. (The signature shown here was written in August 1970, when Zollinger attended ISNA-1 in Sendai as the representative of IUPAC's Organic Chemistry Division.) When Zollinger travels to Japan he carries in his pocket his self-made German–Japanese dictionary, and he likes to speak Japanese during his visits (of which there have been more than nine).

In 1989, ISNA-6 was held in Osaka, with Professor Ichiro Murata as organizer. Ichiro Murata is a modest and sincere person. He decided to enter Tohoku University after reading a review[39] I had written about our hinokitiol studies in Taiwan. He was still in high school at that time. His postdoctoral stay with Professor R. Breslow was one factor that led to the Sendai ISNA meeting mentioned earlier. Professor Murata succeeded in the syntheses of the valence isomers of azulene and heptalene. He is currently involved in the molecular design of novel π-electron redox systems.

Professor K. Hafner of Technische Hochschule in Darmstadt talks with Yoshio Kitahara and Tetsuo Nozoe after Hafner's plenary lecture at ISNA-1 in Sendai in August 1970.

At the ISNA-1 banquet in Sendai, August 1970. From left: Professor H. C. Brown (Nobel Prize, 1979), T. Shimano (mayor of Sendai), Tetsuo Nozoe, and Professor R. Breslow of Columbia University.

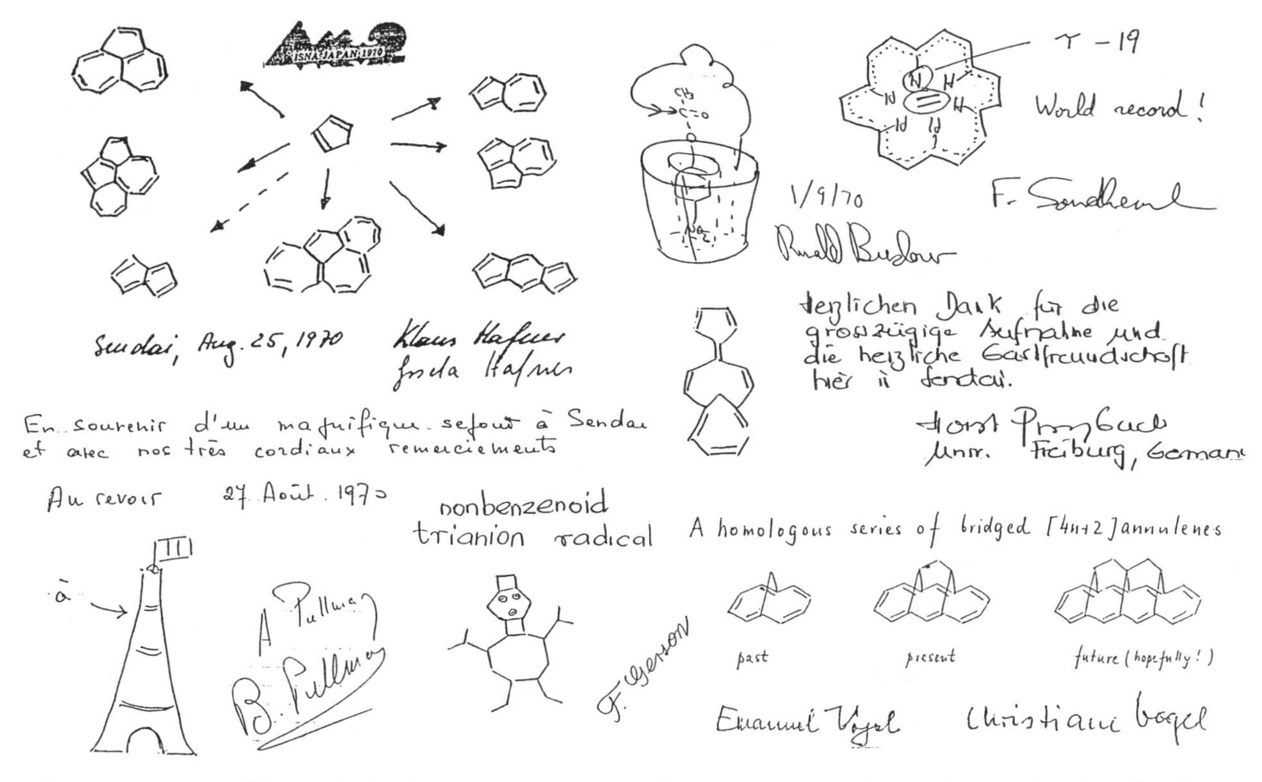

Plenary lecture at ISNA-1 in Sendai, August 1970. ***Professors Hafner, Pullman, Gerson, Breslow, Sondheimer, Prinzbach, and Vogel.***

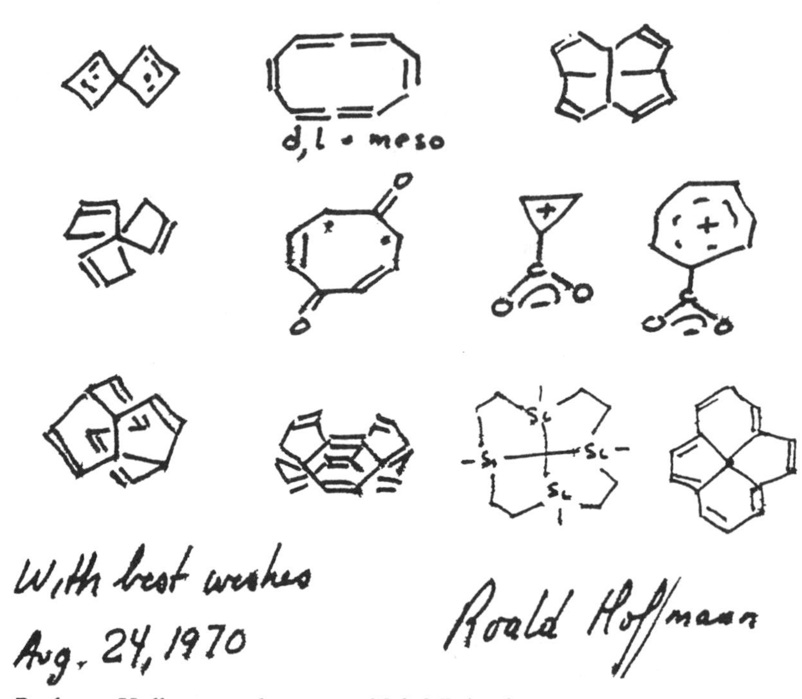

Professor Hoffmann, who won a Nobel Prize in 1981, was one of the plenary lecturers at ISNA-1 in Sendai, August 1970.

Joint Research with Other Groups. For the benefit of the research and development of troponoid chemistry, I had been nursing the idea of building laboratories to be used by our visiting chemists. However, in the early 1960s, poor financial circumstances made it difficult to obtain donations from industrial companies. Therefore, in 1965, I decided to donate my retirement allowance for my 40-year service as a public official toward the building of small laboratories at Tohoku University. This gesture was also my expression of gratitude to the various parties who had, in the past, been of great assistance to me. These laboratories were completed just before my retirement. I myself borrowed these rooms as the first "visiting professor". There, I intended to supplement our studies made thus far and to publish them upon their completion.

At about that time, Sir Robert Robinson visited me in Sendai. He told me that there were now so many contributions from Japan to *Tetrahedron* and *Tetrahedron Letters* that he wanted me to be a regional

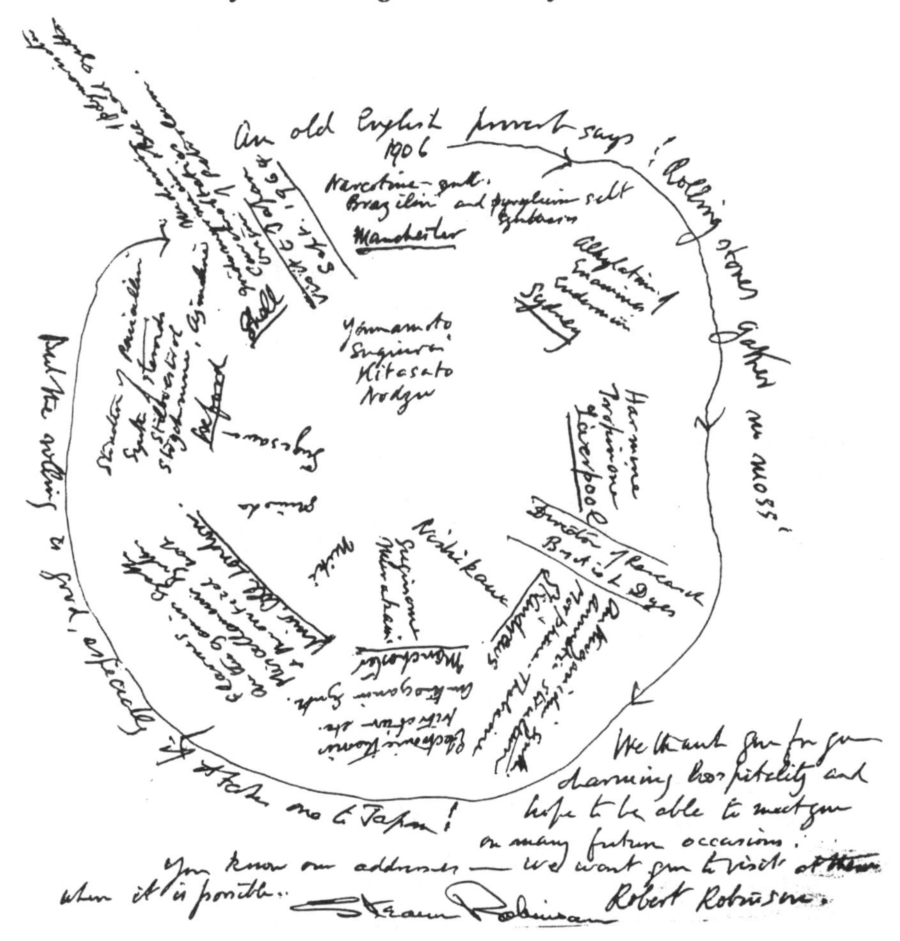

Sir Robert Robinson (Nobel Prize, 1947) visited Sendai in 1964. He spent a lot of time in his hotel room listing the names of all his Japanese students, his personal history, and the titles of research subjects he had studied at various universities or companies.

editor. I accepted the position, although I was not experienced in this kind of work. The following 2 years were very busy and passed quickly, with the editorial responsibilities forcing me to leave my own research unfinished. At about that time, my friend Professor K. Tsuda retired from Tokyo University and became president of a private pharmaceutical college. Because I thought he was more highly qualified to fill the position of regional editor, I asked him to take my place. To my relief, he accepted the position, and in 1968, I moved to Tokyo, because the usual practice in Japan was that retired professors should not remain at the same university. Unfortunately, the troponoid laboratories that I had dedicated to Tohoku University disappeared when the chemistry department moved to a new campus.

At the request of my long-time friend, Dr. Kawakami, Dr. Yoshio Maruta, president of the Kao Soap Company (later called the Kao Corporation), kindly provided me with facilities in the Kao research laboratories in Tokyo. I was to become the research advisor for the company and to have a young assistant, whom they asked me to train in basic organic chemistry. Furthermore, they even provided me with a secretary so that I would be able to carry out my work. My ambition was to begin the preparation of a comprehensive monograph on troponoid chemistry in English. This project had been in my mind for 20 years. Since 1970, my assistants, all non-organic chemists, have helped me continue the research on troponoids. Because I trained a new assistant every year and because this laboratory is not a chemical institute, it would have been unreasonable to expect new discovery from non-organic chemists who were dealing with troponoids for the very first time. Therefore, I placed much emphasis on training through experimentation in tropolone chemistry rather than on research itself. This training method has been a success, and the company, as well as my assistants, have appreciated my system.

As for the reexamination of the unprecedented and delicate points in troponoid chemistry, I asked several professors to collaborate with me on my studies. One of my later students, Hiroshi Yamamoto, received his master's degree under my supervision, and then because of my retirement, he went to the Australian National University in Canberra for his Ph.D. under the direction of Professor A. Albert. He then spent 2 years with Professor W. Pfleiderer (University of Konstanz) on a Humboldt fellowship and another 2 years with Satoru Masamune in Edmonton. Upon his return, he became an assistant professor at Okayama University and a full professor in 1985. He is sincere, highly motivated, and writes good English with ease. He remains one of my research consultants, as well as a close co-worker.

In my laboratory at Kao, using HPLC (high-performance liquid chromatography) (with Dr. H. Okai), we checked the reactions reported by the collaborating researchers. Takasago Perfumery Company and Sankyo Pharmaceutical Company also kindly supported me in various aspects of the research.

My partners during my joint research after my retirement, at first, have been mainly small groups of professors who have had no more than one graduate student. Hence, the research themes only complemented my unfinished work from my days at Tohoku University. I also began investigating interesting themes studied by my former co-workers that had been suspended halfway. I was afraid to see this new information forgotten. The progress of these investigations was natural-

Final lecture at Tetsuo Nozoe's retirement ceremony at Tohoku University, Sendai, in 1966. (Photo courtesy of K. Nakanishi.)

Three long-time friends, enjoying retirement, meet at the Kao Club in Tokyo in 1971. From left: K. Iwamoto (professor emeritus at Gumma University), Dr. Y. Kawakami (president, Kawakami Fine Chemical Corporation), and Tetsuo Nozoe (professor emeritus at Tohoku University).

An excursion to Lake Ashinoko, Hakone, after the close of the 9th International Congress of Heterocyclic Chemistry in Tokyo, August 1983. From left: Professor W. Pfleiderer of the University of Konstanz, Tetsuo Nozoe, and R. Neidlein of the University of Heidelberg.

ly slow, because the situation was completely different from my past experience with large research groups. As my collaborators were all based far from Tokyo, I had to contact them on the phone or by mail (recently also by telefax) at least once a week. A small-scale research project that I would intend to finish in half a year or so frequently required more than 5 years to complete. However, these delays were mainly a result of new occurrences, such as unexpected reactivities, discovered one after another. Fortunately my group of co-workers gradually increased to include distinguished organic chemists. I shall now briefly describe these joint studies.

Extension of Azulene Chemistry

Mechanism of Our Azulene Synthesis. The mechanism of the azulene synthesis from the reactive troponoids, **86**, and the active methylene compounds (AMCs) was reinvestigated. I had previously assumed that the reaction, which was called the Nozoe azulene synthesis,[157] would proceed via four principal intermediates, **A–D** (Scheme VIII).[90,151,156] Although there was no direct proof of the existence of intermediates

B–D, the mechanism was considered to be reasonable at that time. However, the fact that trisubstituted azulenes, **127**, were usually produced very rapidly and in very good yields especially from the imine **122a** (R = CN, X = NH) and **125a** or **125c**, it became inconceivable to me that the reaction would proceed through several such stable intermediates, as shown in Scheme VIII (p 99).[158]

First, I considered it necessary to prove the existence of the forked heptafulvene **C** and azulenone intermediate **D**. I expected that the intermediate **C** or **D** might be obtained by the reaction of **122a** and **125d** or **125b** because of the difficulty in producing the final product azulene in these cases (*see* Table II). Shiro Kawahito and Akio Kimura were able to isolate compounds having the same compositions as forked heptafulvenes, **C**. However, by NMR spectroscopy, we found that these compounds were Michael addition products of **122a** with **125d** or **125b** at C-6, C-4, or both (e.g., **131a** and **131b**[159] in Scheme IX). Therefore, I suggested that the reaction should be examined spectrometrically from the beginning to the end. This idea stemmed from my experience in Formosa more than 50 years before, when I had managed to follow the color reaction of sugars by using a manual spectrophotometric method[17] (*see* the earlier section, Saponins and Sapogenins).

Fortunately the next co-worker who came to my Kao laboratories was an analytical chemist, Jiro Kawase. He constructed an HPLC apparatus equipped with a stopped-flow UV–visible spectrophotometer and periodically checked various reactions by the combination of substrate (**86** or **122**), solvent (CH_3OH, CH_3CN, THF [tetrahydrofuran], etc.), and base ($NaOCH_3$, *tert*-butylamine, pyrrolidine, etc.). As a result, a surprising number of competing reactions (reversible and irreversible) was observed in the solution from the beginning to the end.[160] The reactions of a methanolic solution of **122a** and **125d** in the presence of base (*tert*-butylamine) at room temperature were followed by reverse-phase HPLC (Figure 3). Three small peaks, c_1–c_3, were observed instantly on the addition of **125d** to a methanolic solution of the imine **122a**. Among these peaks, c_1 and c_2 were due to the adducts **131a** (at C-6) and **131b** (at C-4) mentioned previously, whereas c_3 was an adduct, **131c**, of solvent (methanol) at the 2-imino group.

The addition of *tert*-butylamine as the catalyst to this solution instantaneously resulted in the appearance of two large peaks, b_1 and b_2, due to the Michael addition products from the addition of amine at C-4 and C-6 (**131d** and **131e**). Almost no change was observed after the solution was set aside overnight at room temperature. However, 22 h later, and especially upon refluxing, a total of 13 peaks appeared, as shown in Figure 3. Each of the peaks was characterized by means of retention time, UV spectroscopy, and mass spectrometry, as well as R_f

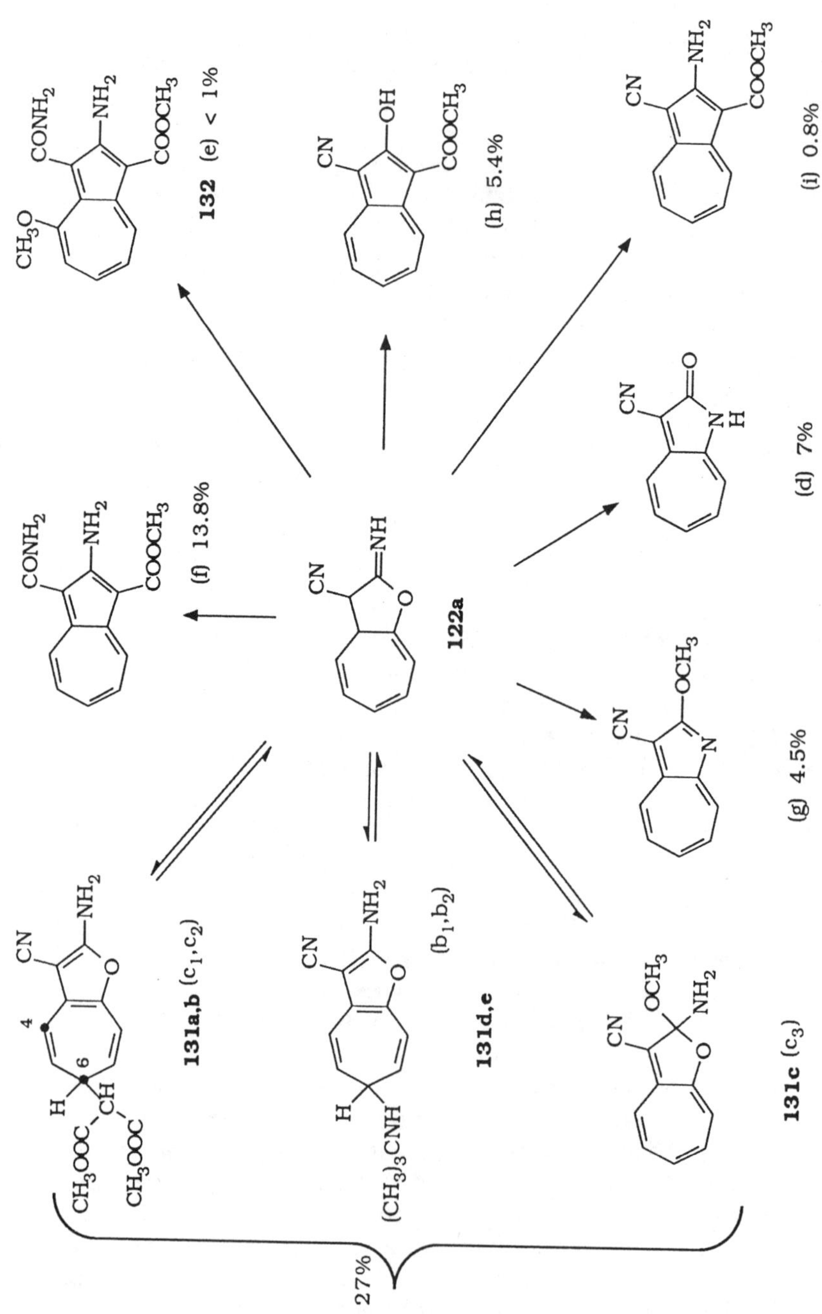

132 (e) < 1%

(h) 5.4%

(i) 0.8%

(f) 13.8%

122a

(d) 7%

(g) 4.5%

131a,b (c_1,c_2)

131d,e (b_1,b_2)

131c (c_3)

27%

Scheme IX

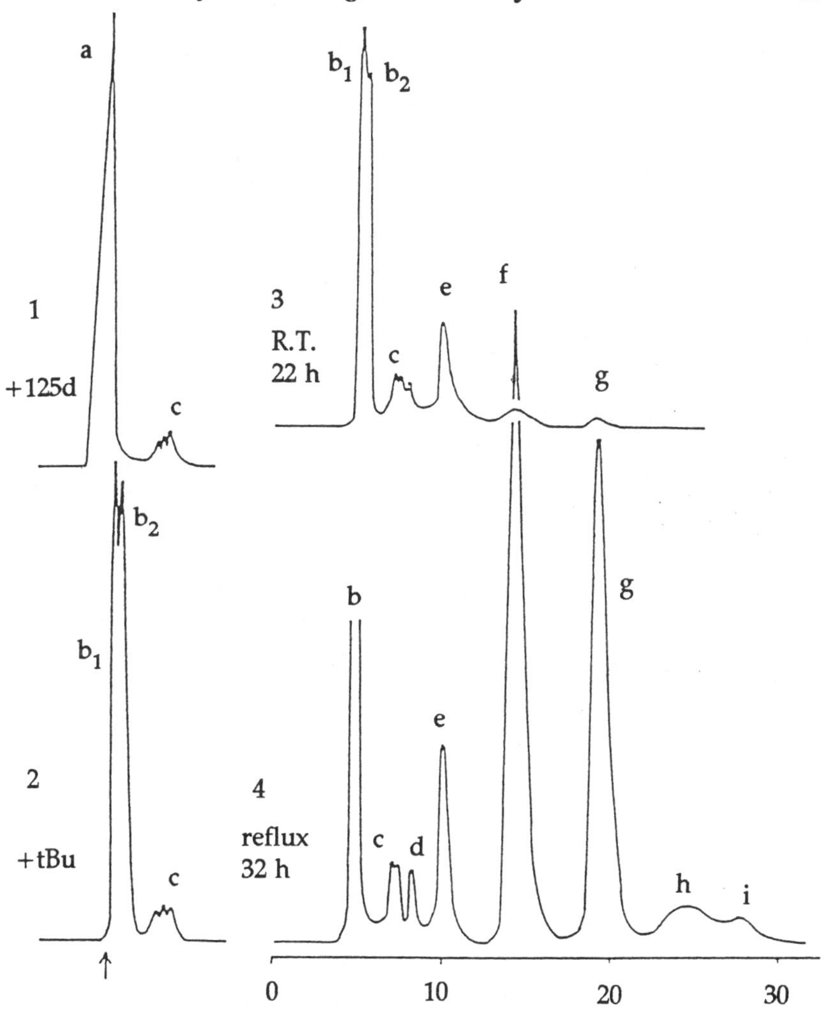

*Figure 3. Results of the time-dependent HPLC analysis of the reaction of **122a** with **125** (R = CH_3) in methanol in the presence of* tert-*butylamine. R.T. is room temperature.*

values (in thin-layer chromatography [TLC]). The formation of the compounds (together with approximate yields) is illustrated in Scheme IX. Heating of the solution mixture probably caused loss of the specificity of the reaction paths. This loss of specificity explains the formation of various azulenes through both paths a and b.[160] Interestingly, a 4-methoxyazulene derivative, **132**, was found in peak e, which is believed to have been produced by the autoxidation of the adduct formed by the addition of methanol to **122a** at C-4. The result clearly indicated the competitive attacks at various positions of the substrate **122a**, not only by the reagent (**125d**) but also by the base and solvent.

On checking the reaction of **122** (**A**) with AMC, as described in Table II, by time-dependent HPLC, it was generally observed that azulenes are produced very fast and in very high yields, although the tentative formation of the Michael reaction adducts (**131a–131d**) can sometimes be seen as the only detectable byproducts. Treatment of **122h** with **125a** at room temperature easily gave azulenopyridone **128**[154] in 91–98% yield regardless of the kind of solvent (CH_3OH or CH_3CN) and base (*tert*-butylamine or $NaOCH_3$) used (Scheme VII), and no intermediates (**C** and **D**) were observed by HPLC.[161] These facts led us to conclude that the azulenes **127** are formed as a result of consecutive reactions through a series of metastable anionoid intermediates (B^-, C^-, and D^-), as I had predicted previously.[158]

Azulene Synthesis with a Vinyl-Ether-Type Reagent. Professor Takase and his co-workers have found[162] a convenient synthetic method for the various azulenes **127** by the reaction of the same substrate, **122e** (R = CO_2CH_3, X = O) with enamines or its precursors (aldehyde and morpholine). (Professor Kahei Takase at Tohoku University (with Dr. K. Takahashi, M. Yasunami, and O. Morita) and Professor Hitoshi Takeshita at Kyushu University, all my former students, are currently considered as central figures in troponoid chemistry in Japan.)

Meanwhile, with Professor P.-W. Yang (my former student, who is now at the National Taiwan University), Professor Ishikawa, H. Wakabayashi (Josai University), and Kao Tokyo Research Laboratories (with Dr. H. Okai), I established a versatile, one-pot azulene synthesis by using an [8+2] cycloaddition of the troponoid precursors **122**. When **122** (R = H, $COOCH_3$, $COCH_3$, CN, or C_6H_5; X = O) is heated with vinyl ethyl ether at 150–200 °C for 10–40 h in a sealed tube, the azulenes **127** were easily obtained in one pot in a 40–90% yield.[163]

By HPLC examination, a competitive formation of the [4+2] cycloadducts (C_1 and its isomers) on the seven-membered ring (three isomers were determined by NMR techniques) was observed in addition to the azulenes. The [8+2] cycloaddition product **A** initially formed as a main product is believed to easily yield azulene by the elimination of CO_2 (**B**) and then ethanol, as shown in Scheme X. The use of vinyl acetate or propenyl acetate was found also to give azulene derivatives but in low yields; in this case the [4+2] cycloadducts were the main products.[163] By using dihydropyran, dihydrofuran, and furan derivatives, a wide variety of azulenes containing versatile functional groups on five-membered ring became available in one pot and in high yields

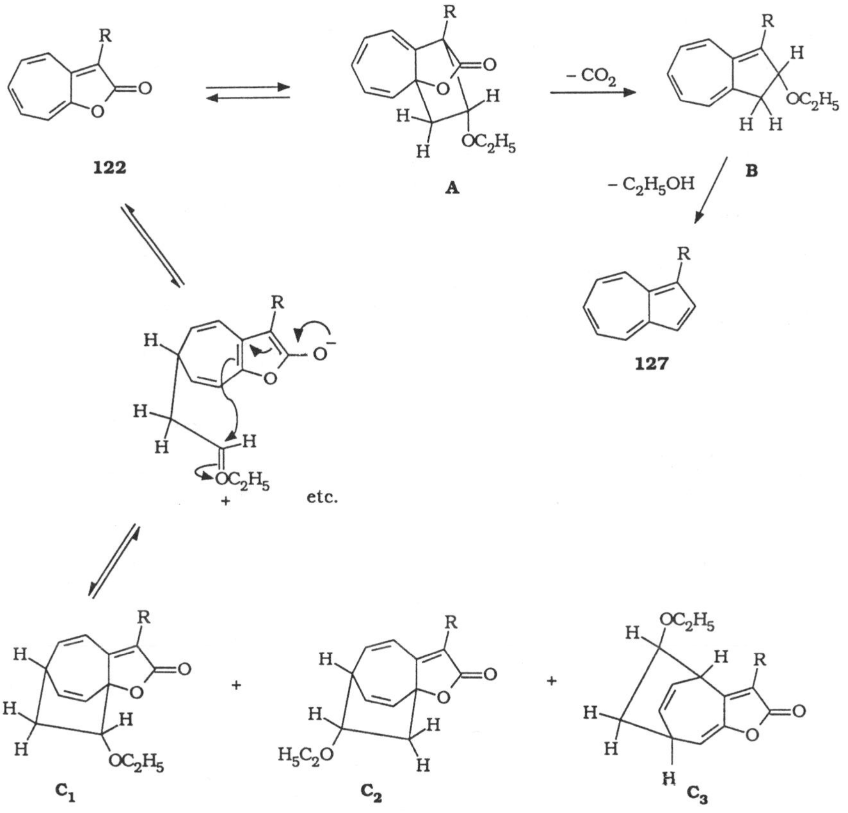

Scheme X

(Scheme XI).[164a] When substituents on a seven-membered ring are desired, an appropriately substituted tropolone can be used accordingly as the starting material. This method could be easily extended to use aldehydes, ketones (as acetals),[164a] and 2-alkoxyazulenes, which are directly produced in high yield by using orthoacid esters as reagents.[164b]

Autoxidation of Azulenes. Azulenes have been regarded as one of the representative examples of nonbenzenoid aromatic hydrocarbons, which usually do not undergo Diels–Alder-type cycloaddition reactions but are easily susceptible to many electrophilic substitutions. Guaiazulene (133), which is easily prepared by the dehydrogenation of guaiol, a constituent of guaiac wood, has long been known to change to various colorful substances during storage, although the precise structures of these autoxidation products had remained virtually unestablished. Also, sodium guaiazulene-3-sulfonate, which is increasingly used clinically as

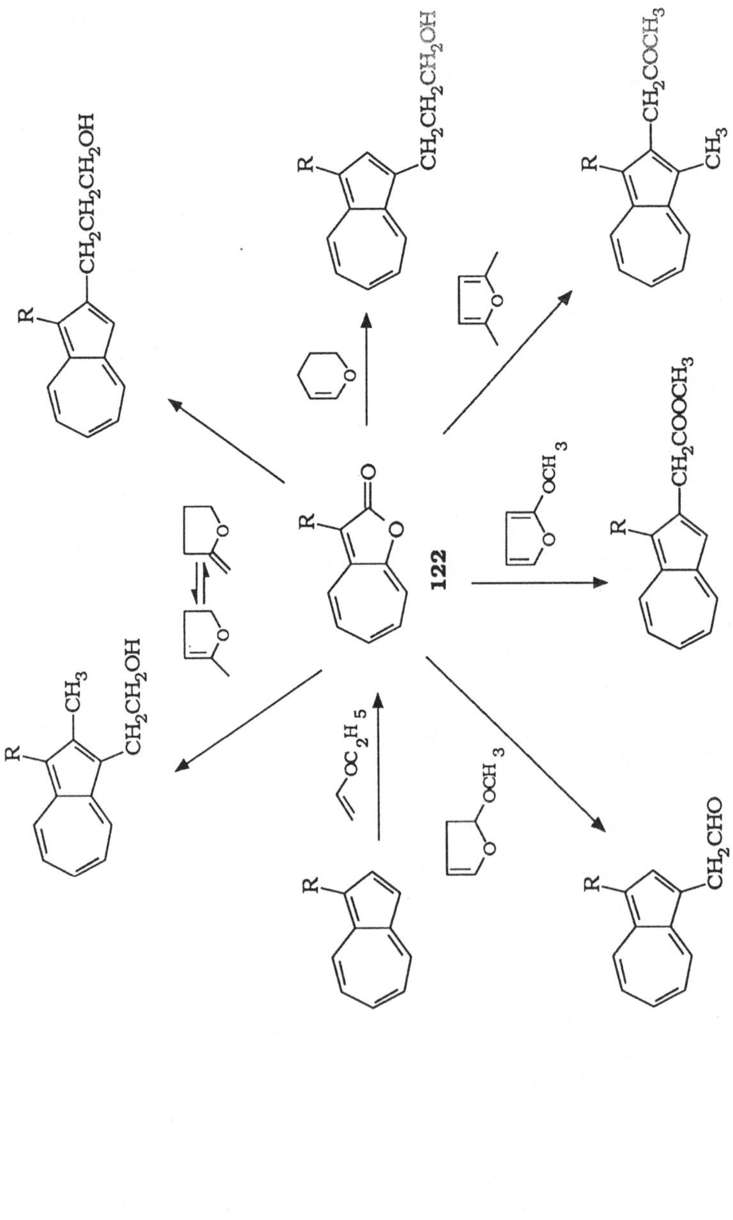

Scheme XI

an anti-inflammatory drug in recent years, has been noticed to change slowly into unidentified products during storage. On the other hand, various azulene derivatives having an electron-withdrawing group at C-1, C-3, or both, which were synthesized by our method, have remained unchanged for more than 30 years. A striking contrast between these products made us closely study the autoxidation (as well as peracetic acid and hydrogen peroxide oxidation) of azulenes, with guaiazulene as the first model compound because it has a high redox potential differing from other azulenes and is easily oxidized even at room temperature. These studies were carried out with Professors Yoshiharu Matsubara (Kinki University, Higashi-Osaka) and Professor Hiroshi Yamamoto (Okayama University).

The autoxidation reaction of guaiazulene (**133**), for example, in DMF (dimethylformamide) at 100 °C, was checked by means of HPLC and TLC (Figure 4). We found that the oxidation was extremely complex, and many products appeared simultaneously right from the beginning. After careful separation of the reaction products by silica gel column chromatography, TLC, and reverse-phase HPLC, more than 30 compounds having highly interesting structures were isolated, representative examples of which are shown in Chart IV.[165] These compounds include a side-chain oxidation product (**134**), azulenequinone (**135**), naphthoquinones (**136** and **137**), indenones (**138** and **139**), a benzenoid compound (**140**), 3-formylguaiazulene (**141**), oxidized 3,3′-coupling products (**142**), 3,3′-methylenebis(guaiazulene) (**143**), a ketone (**144**), and others.[165]

Compounds having an extra carbon atom, such as **141**, **143**, and **144** are of particular interest. Similar products were always isolated by the autoxidation of azulene itself and other alkyl azulene derivatives.[166] As important intermediates for such intermolecular one-carbon-transfer reactions, 6-(3-guaiazulenyl)-(6*H*)-guaiazulenone (**145**) and its norcaradiene isomer (**146**) were isolated. Incidentally, for this type of valence bond isomers, this was the first time that each compound became available in a pure form, as dark-green (**145**) and blue (**146**) prisms.[167] These compounds exist in a temperature-dependent equilibrium at −20–60 °C in oxygen-free hexane. Then, **146** was found to decompose quantitatively into 3-formylguaiazulene (**141**) and indenone (**139**) when dissolved in an organic solvent (e.g., DMF, CH_3OH, or $CHCl_3$) and exposed to air, whereas under nitrogen in the presence of one equivalent of guaiazulene (**133**) in DMF at 100 °C, **146** gave **143** and **139**. Surprisingly, when we dissolved **145** in $CDCl_3$ to examine its NMR spectrum, we found that **145** instantly decomposed into 3-deuteroguaiazulene and the quinonoid compound **147**. This decomposition must be due to the existence of a trace amount of DCl in the solvent (Scheme XII). Presumable mechanisms of the unprecedented one-carbon transfer during autoxidation of azulenes were clarified by

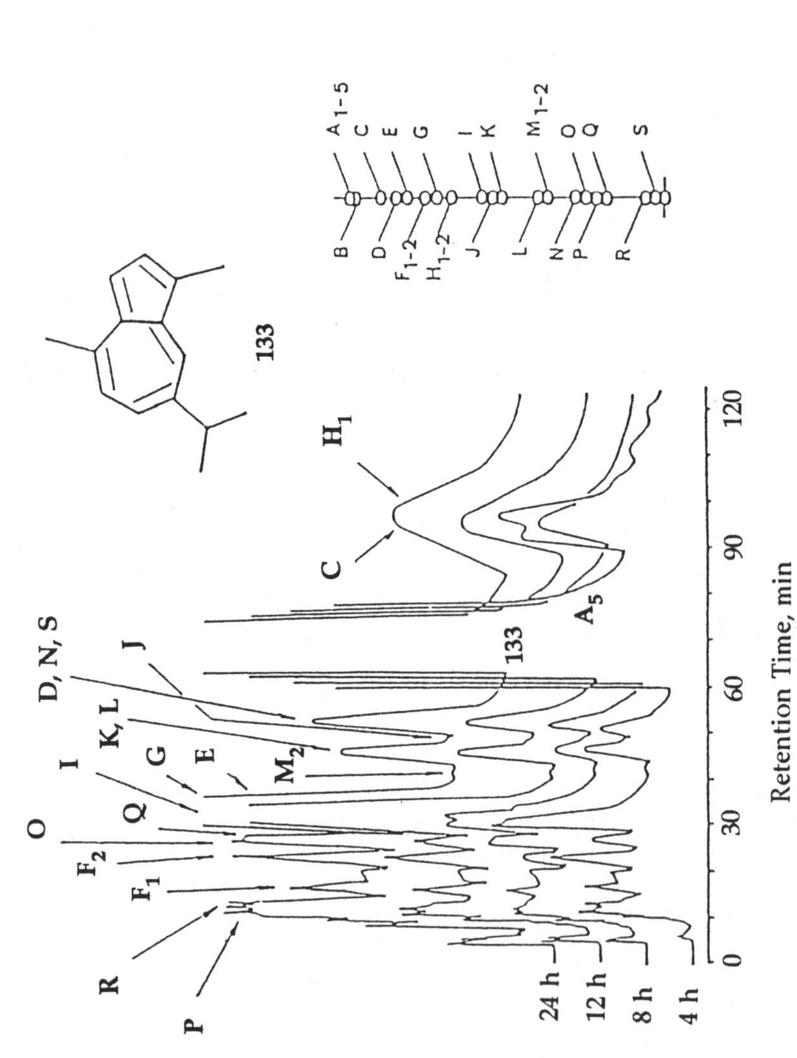

Figure 4. Results of the time-dependent reverse-phase HPLC analysis of the autoxidation of guaiazulene (133) (left) and thin-layer chromatography of the products (right). (Reproduced with permission from ref. 165. Copyright 1984 Chemical Society of Japan.)

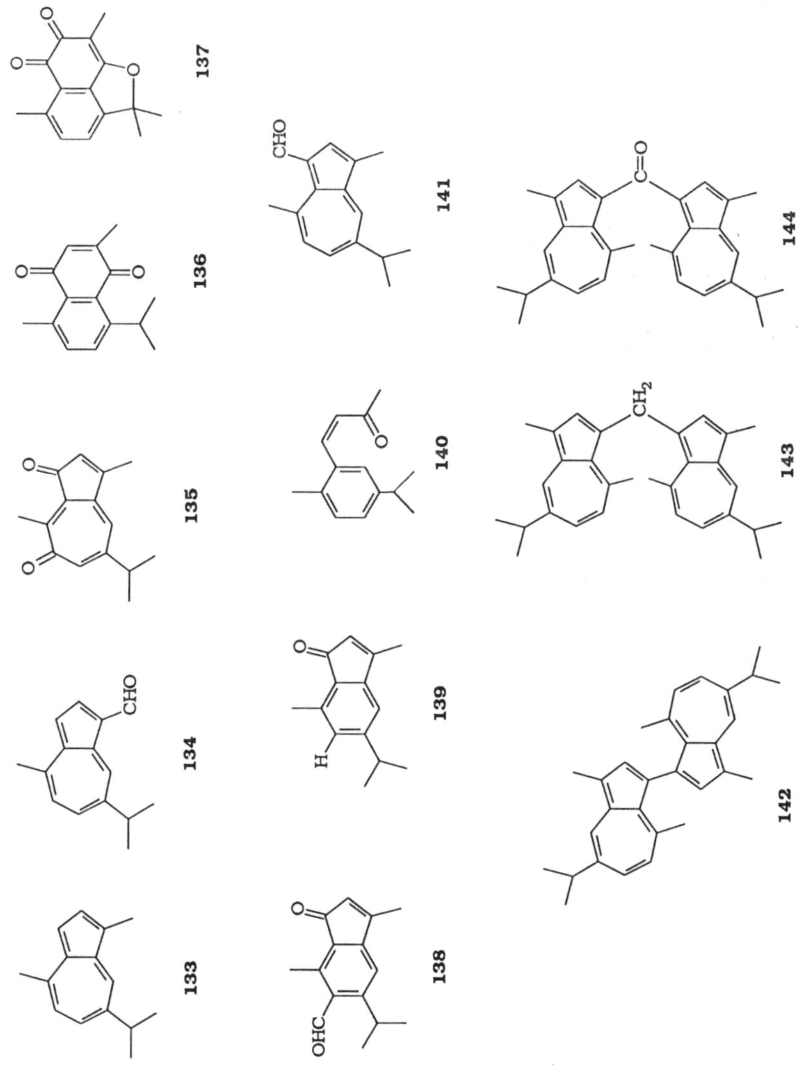

Chart IV

133 134 135 136 137

138 139 140 141

142 143 144

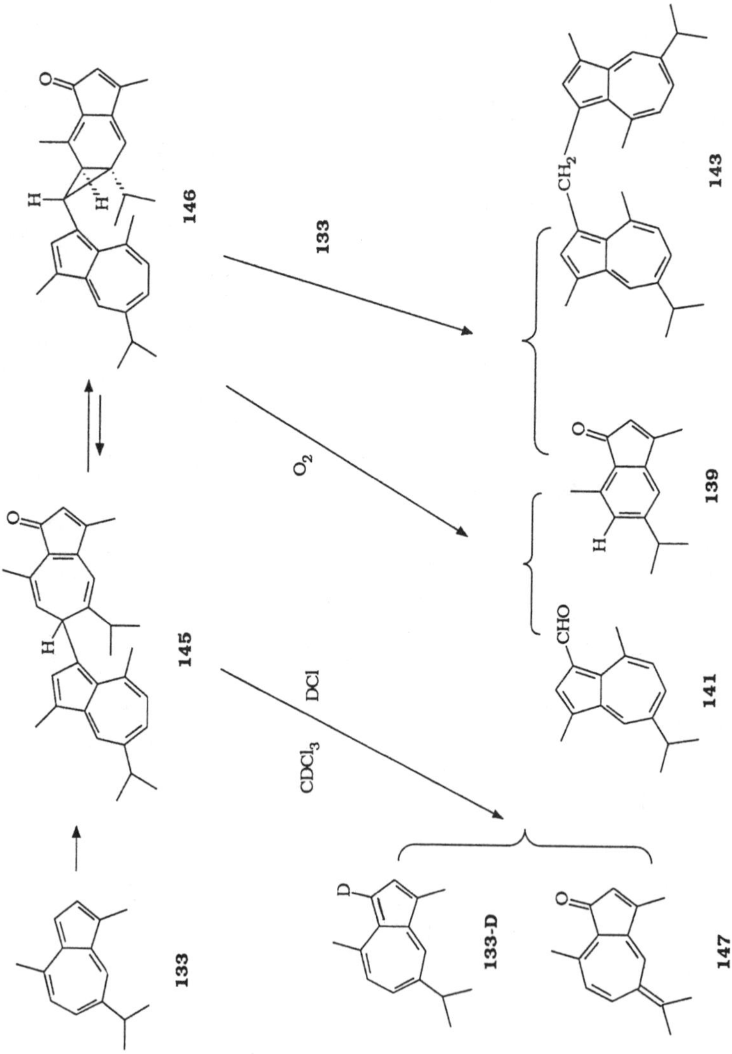

Scheme XII

the isolations of guaiazularyl radical as dimers by the peracetic acid oxidation of guaiazulene.[168,169]

We have found recently that on allowing guaiazulene (**133**) to stand in the dark for several months at room temperature, guaiazulene-quinone (**135**) is slowly formed as one of the main products (10–15% yield after isolation),[168] whereas treatment of **133** with 30% peracetic acid in acetic acid rapidly yields **135**, **145**, and **146** in 20, 20, and 18% yields, respectively.[169] This result is very interesting in relation to the recent report by Scheuer and his co-workers[170] on the isolation, from blue polyps of a deep-sea gorgonian (at –350 m), of **133**, **135**, **141**, and **143**, besides 3-chloroguaiazulene (**148a**, X = Cl), 3-bromoguaiazulene (**148b**, X = Br), ehuazulene (**149**), and artemazulene (**150**) as natural products. It appears to me that all of these compounds are derived from **133** by enzymatic oxidation in the deep sea water. Therefore, it has been a pleasant surprise to me to find out also recently that treatment of **133** with NBS in cyclohexane under argon exclusively produces 3-bromoguaiazulene (**148b**), whereas the same reaction in benzene in open air yields a complex mixture of **148b**, **149**, **151**, **152a** and **152b** (geometrical isomers), and **153** (Chart V), in addition to some structurally unestablished products.[171]

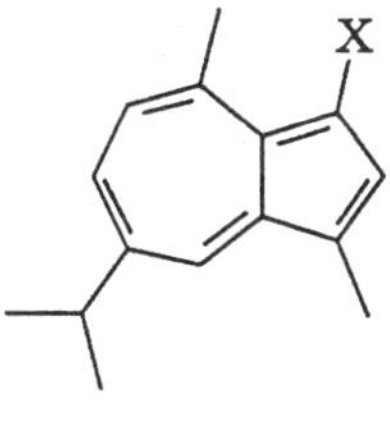

148a,b 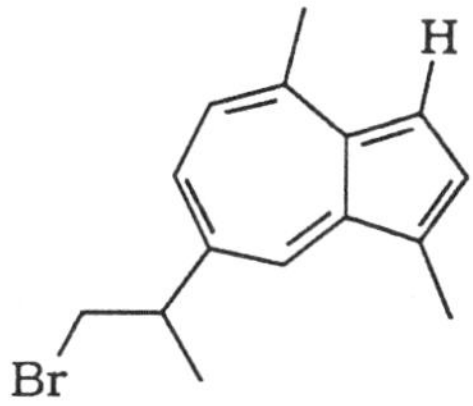 **149** 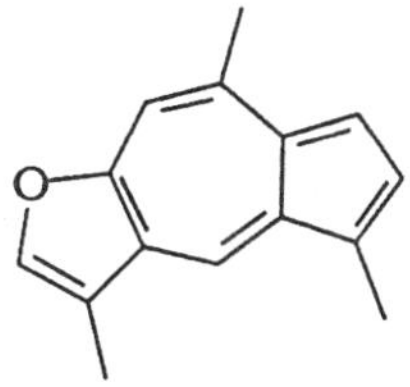 **150**

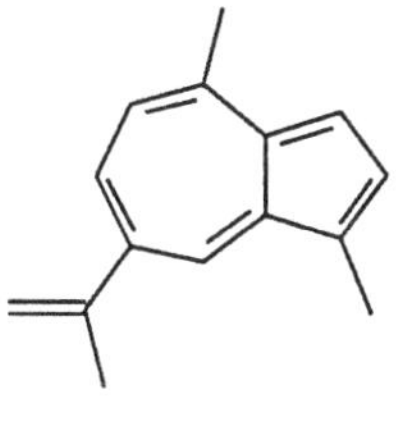

151 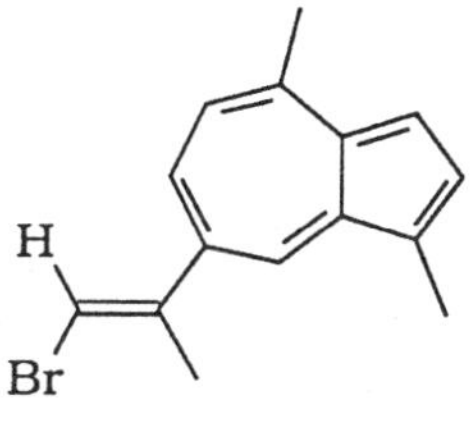 **152** 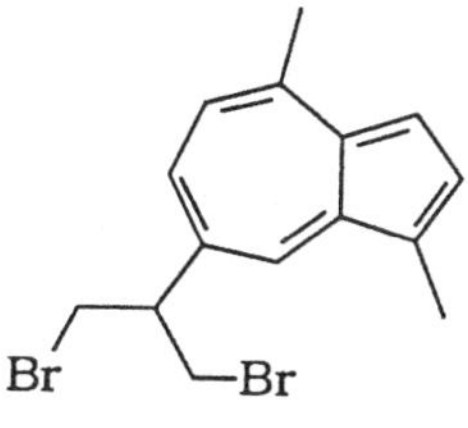 **153**

Chart V

Some Topics in Tropolone Chemistry

Ring Opening of Tropone Oxime Tosylate. In 1965, Professor Kitahara and Makoto Funamizu found a facile ring-opening reaction of tropone oxime tosylate (**154**, $R^1 = R^2 = H$) but gave up the study uncompleted. With Professor Takahisa Machiguchi (Saitama University), I have recently undertaken a careful reinvestigation of a novel ring opening of tropones (**154**) by using various kinds of nucleophiles. For example, as shown in Scheme XIII, when pyrrolidine is used as a nucleophile, the reaction easily proceeded in CH_2Cl_2 at −20 °C within 1 h and initially gave the all-(Z) product **155** (Nu = 1-oxa-4-azacyclohexane, $R^1 = R^2 = H$) in 97% yield, which isomerized to the (Z,Z,*E*) compound **156** in the presence of trace of acid. Compound **156** was then converted almost quantitatively to the all-(*E*) form **157** in an alumina column or in a more potent acid. It was confirmed that the reaction with other nucleophiles also took place easily, but the structures and stability of the trienes depended on the type of nucleophile used.[172] This ring opening is initiated by the attack of a nucleophile at the α position of **154** and followed by the ring cleavage.

Synthesis of B-Ring-Opened Colchicine Analogues. Thirty-five years ago we synthesized simple tropolones, such as **158**[173] and **159**,[174] which showed some interesting biological properties. These compounds were

TsO N R2 R1 NC Nu R2 R1

155

154

Ts = tolylsulfonyl, $CH_3C_6H_4SO_2$

NC R2 R1 Nu NC R1 R2 Nu

157 156

Scheme XIII

active as mitotic poisons. Then we intended to synthesize the bicyclic colchicine analogues (**163**) by the condensation of various benzaldehyde derivatives (**161**) with 4-acetyltropolone (**160**). We obtained the unsaturated ketones **162**, which were then converted to **163** by the reductive

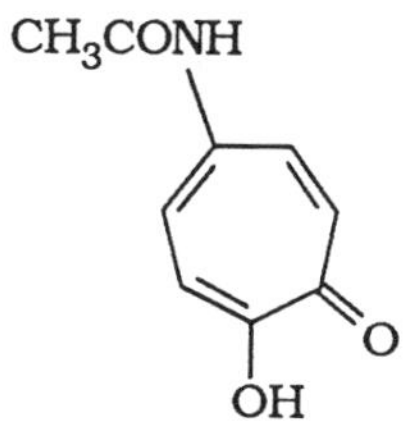

158 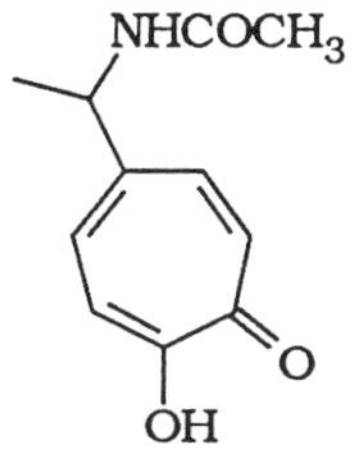 **159** 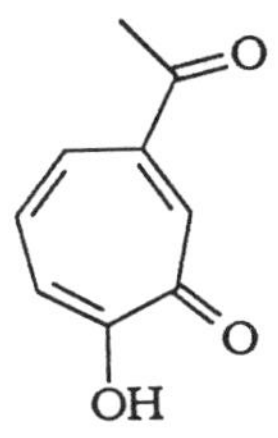 **160**

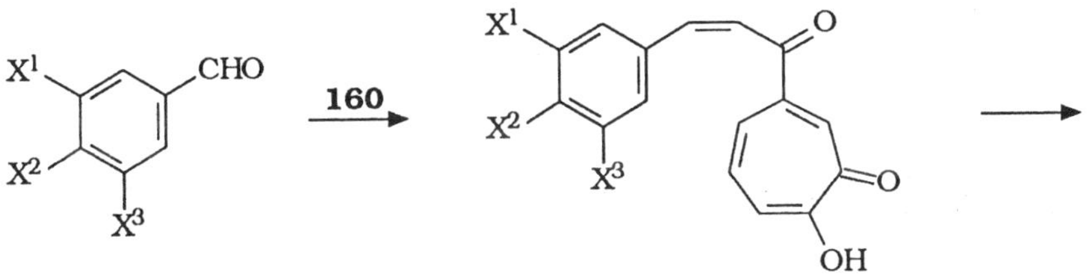

161 **162**

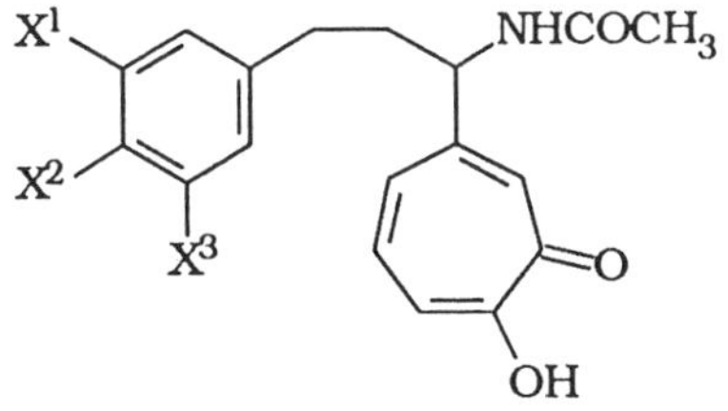

163

acetylation of their oximes.[175–177] However, we did not succeed in synthesizing colchicine (**123**) from the 5-amino derivative of **163**.[175]

We recently started to develop a novel synthesis of colchicine analogues of the type **165**, which, we believed, has a conformation similar to that of the natural colchicine **123**. Although the tropolone nucleus is highly susceptible to various electrophilic substitutions, it does not undergo Friedel–Crafts-type alkylation or acylation. The reaction of organolithium and magnesium reagents with tropolones is known to yield C-3 and C-7 substitution products. Thus, virtually no efficient synthetic route has been established for the preparation of 5-

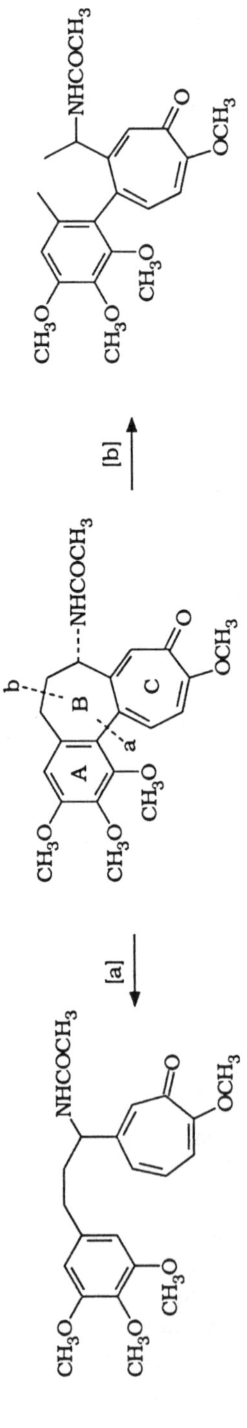

164 [a] 123 [b] 165

aryltropolones. Dr. Yukio Sugimura of Sankyo Company in Tokyo found that treatment of **166** with dilute HCl readily gave a product presumed to be formed through a benzidine-like rearrangement.

Dr. Sugimura had to interrupt this study because of other more-urgent investigations of the company, and I asked Professor Takase and his co-workers (Tohoku University) to study the acid-catalyzed rearrangement of **166** with dilute hydrochloric acid. They found that treatment of **166** (R^1 = H, R^2 and R^3 = H or CH_3) with 2 M hydrochloric acid in ethanol under reflux gave 2-amino-5-(4-aminophenyl)tropones, **167** (R^1 = H, R^2 and R^3 = H or CH_3) apparently by the 5,*p*-rearrangement similar to the usual *p,p'*-benzidine rearrangement.[178] Compound **167** was then converted to the corresponding 5-aryltropolones, **168**. However, Professor Takase's group could not con-

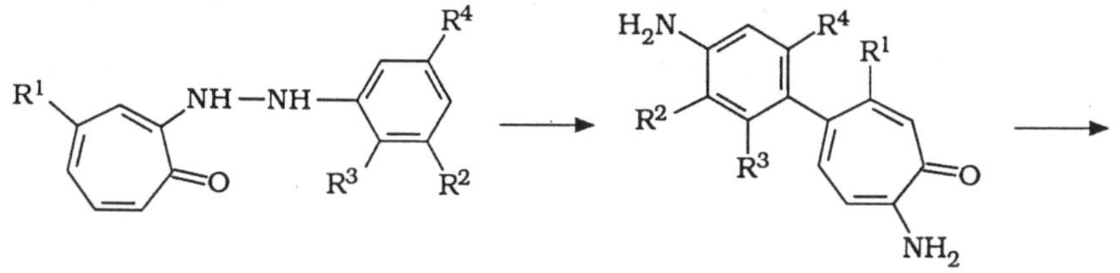

166 **167**

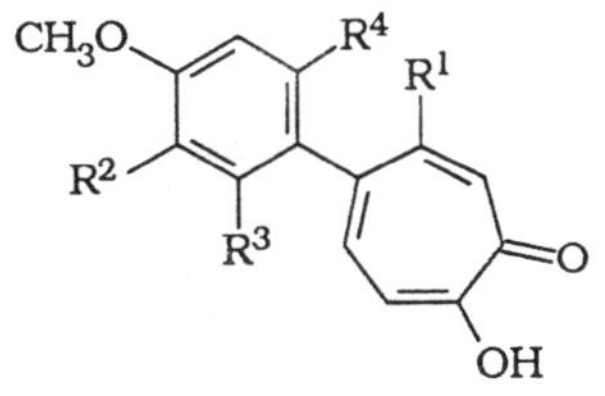

168

tinue this study because of commitments to other projects, and I asked Professors H. Yamamoto and K. Imafuku (Kumamoto University) to prepare 5-aryltropolones having various substituents on both benzene and tropolone rings. Compounds **169** and **170** were thus obtained.[179,180]

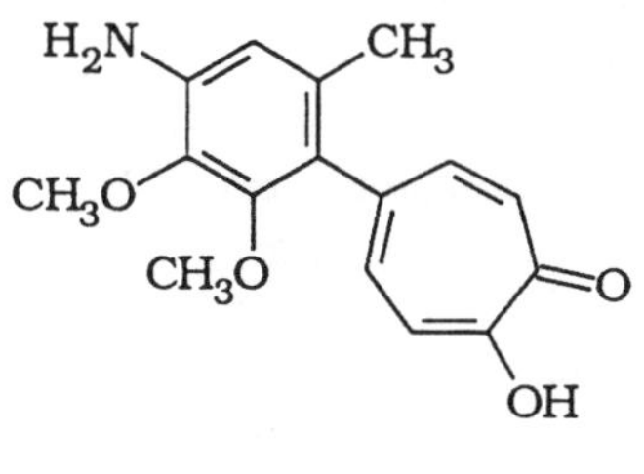

169 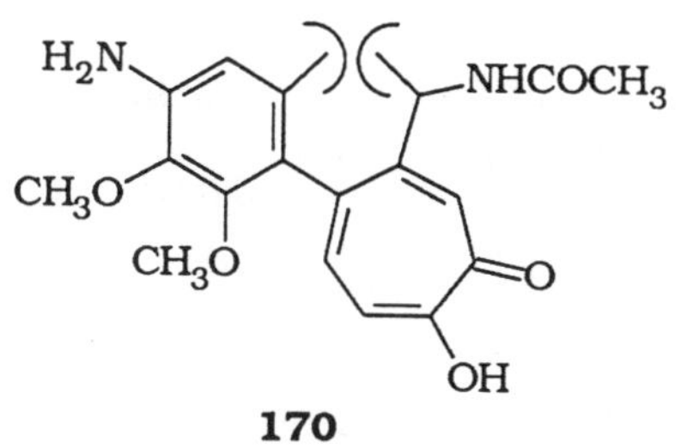 **170**

These compounds are considered to be interesting biochemical tools in obtaining various information on tubulin-binding sites and may also show useful biological activity (e.g., as antitumor agents).

Tropylium Ring Annulated with Heterocyclic Groups

From the synthesis of tropolones and tropones in the early 1950s, a number of troponoids annulated with benzene and heteroaromatic or alicyclic ring(s) have been prepared abroad and in Japan. However, most of these new compounds did not possess the reactivities characteristic of monocyclic reactive troponoids (**86**). In contrast, compounds such as cyclohepta[*b*]furan-2-ones (**122**) and colchicine (**123**) exhibit interesting characteristics due to their 6π-electron tropylium system. Later we found that some tropylium compounds annulated with heterocyclic groups having more than two heteroatoms also exhibited unusual properties because of their dipolar aromatic systems.

Tropocoronands and Their Metal Complexes. A series of new classes of compounds, **171** ($n = 3–6$), **172** (X = O or S), and tristroponeimines, **173**, were recently prepared[181,182] through the kind cooperation of Dr. Imajo (Suntory Institute of Bioorganic Science) and Professor K. Nakanishi (Columbia University). These compounds, which we named "tropocoronands," were synthesized by the condensation of the reactive troponoids with 1,ω-polymethylenediamines or their derivatives in relatively good yields by applying a high-dilution method.[182] Specifically, these compounds complex easily with various metal ions, such as Ni, Cu, and other transition metals. Complexes such as **174** can exist either in the nearly square-planar diamagnetic form, **174a**, or the pseudo-tetrahedral paramagnetic form, **174b**, depending upon the length of the linking chains. Interestingly, a unique sulfonium ion, **175**, containing a hydroxyl anion in the cavity was obtained as a byproduct in the preparation of **172** (X = S) (Chart VI). A detailed NMR study of these tropocoronands and their nickel complexes (**174**) was carried out;[182] meanwhile their stereochemical and electronic spin-state tuning was studied by our colleague, Professor Lippard (MIT).[183]

Cyclohepta[*b*][1,4]benzoxazines. Thirty years ago we believed we had synthesized the parent compounds, **176**, of quinoxalotropone (**75**), which had been named benzo[*b*]tropazines.[184] We also prepared the sulfur analogue, benzo[*b*]tropothiazine[185] (**177**) and its tropone derivatives.[186] Later, Fukunaga[187] reported, however, that he obtained a greenish black, cationic salt, **178a**, which reversibly gave quinoxalotropilidene (**176a**) on neutralization. By comparing the NMR spectrum with those

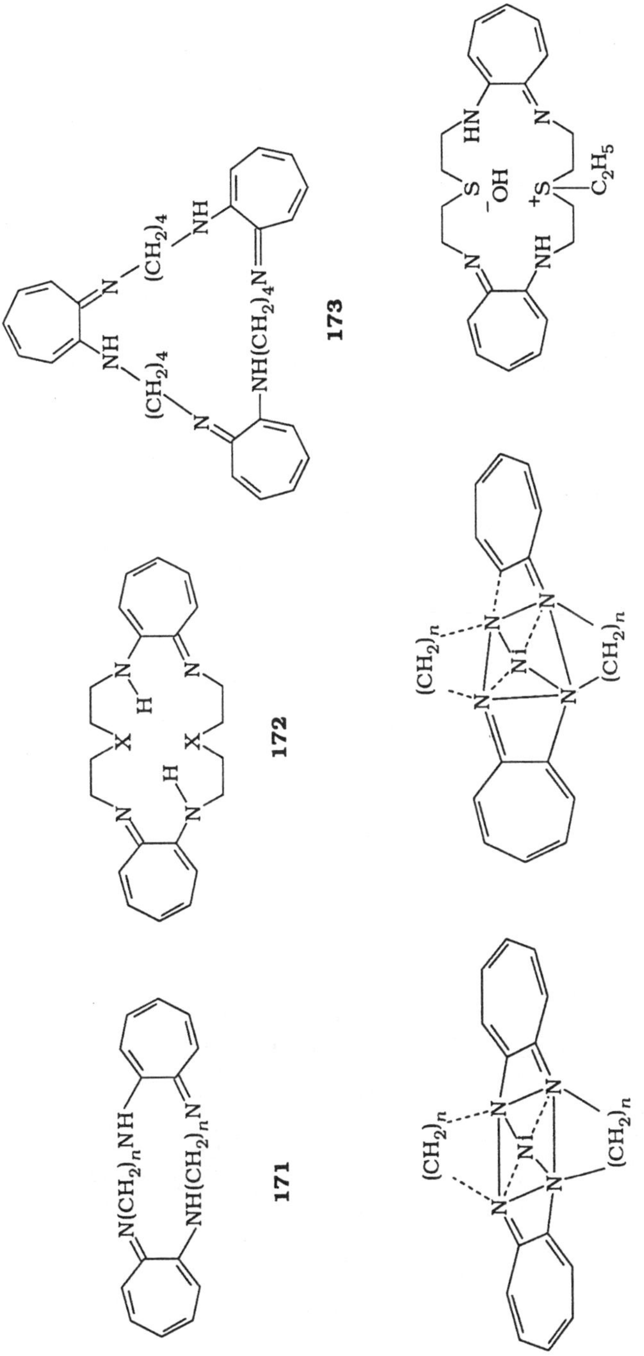

171 **172** **173** **174a** **174b** **175**

Chart VI

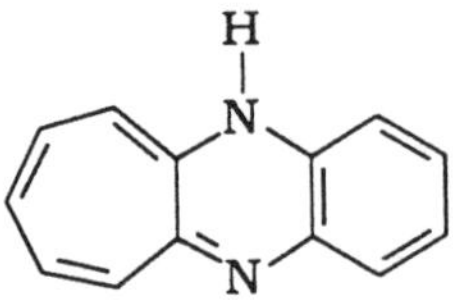

176

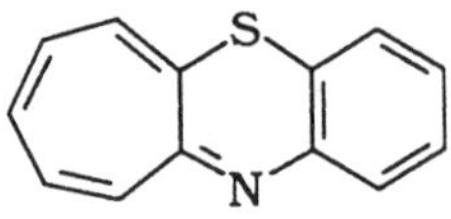

177

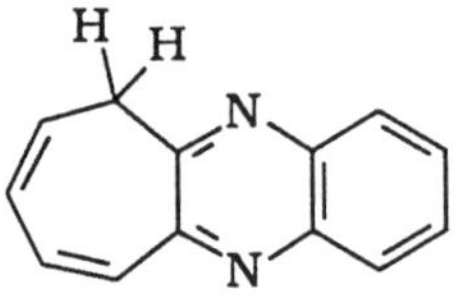

176a

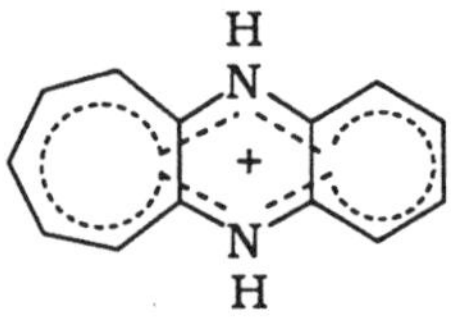

178a

of structurally similar compounds, he concluded that the green cation had the anti-Hückel, 16π-peripheral-electron system.[187] He also found that **176a** was easily autoxidized, especially under basic conditions, and assumed that the white crystals that we had obtained[184] by "purification" through an alumina column were in the form of a dimer, but he did not determine the structure of the dimer. We felt that closer reexamination of these benzotropazine systems was necessary, and thus we examined the reactions of the reactive troponoids (**86**) with *o*-aminophenol (**179**), *o*-aminothiophenol (**180**), or *o*-phenylenediamine (**181**), with the cooperation of the Kao Tokyo Research Laboratories

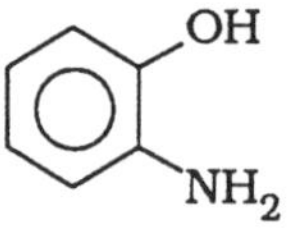

179

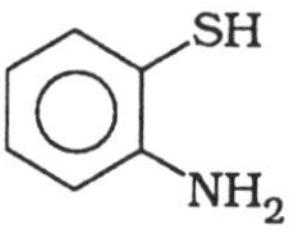

180

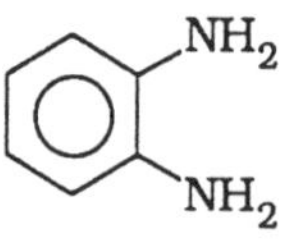

181

(with H. Okai), Professor S. Ishikawa's group at Josai University (with H. Wakabayashi, K. Shindo, and T. Kurihara), and, partly, the Takasago Perfumery Company (with T. Someya). Studies in this area have unexpectedly required a long time and still do.

Cyclohepta[*b*][1,4]benzoxazine (benzotropoxazine), **183**, was easily obtained[188] via 2-(*o*-anilino)tropone, **182a** (X = H), by the condensation of **86** and **179**. Compound **183** was readily hydrolyzed by alkali to generate **182a**, which subsequently gave tropolone and **179** after being heated with excess alkali. Red-colored cations, **184**, formed from **183** in strong acid are stabilized by the delocalized six-π–electron benzenoid and tropylium systems. To confirm the substitution patterns, we examined the reaction products of **179** with 4-, 5- and 6-isopropyl-2-

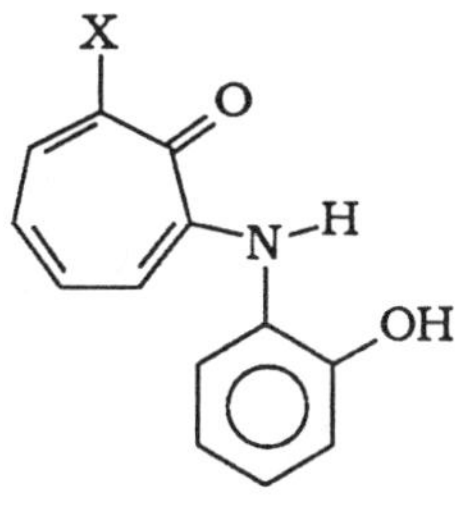

182a,b 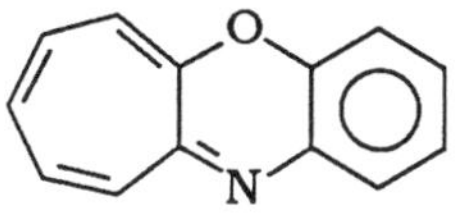 **183** 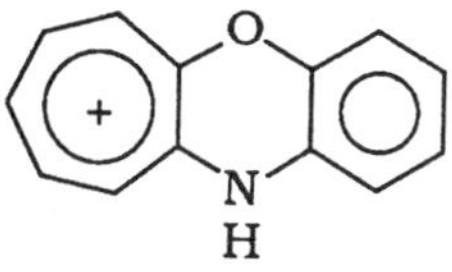 **184**

chlorotropones, **185a–185c**.[189] In all cases, isopropyl derivatives **186a–186c** were obtained in very high yields. Whereas the 5-isopropyl compounds **185b** and **179** provided the C-8-substituted derivative **186b** as a single product, the 4- and 6-isopropyl compounds **185a** and **185c** gave an almost 1:1 mixture of **186a** and **186c**. The reason for this result was not established at that time.

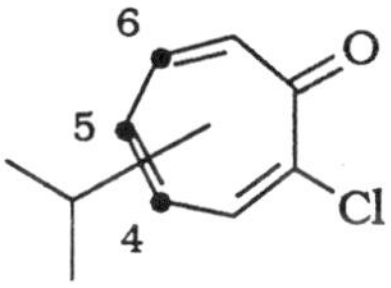

185a-c 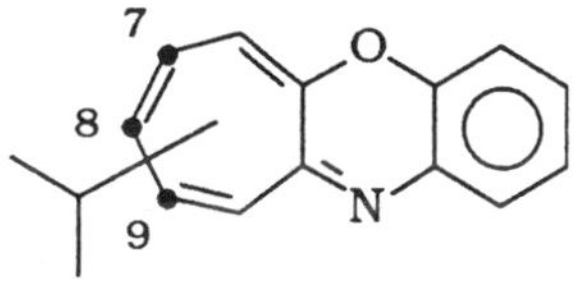 **186a-c**

Then, we compared the reactions of **179** with three isomeric bromomethoxytropones, **87–89**, and encountered an unexpected result, which sometimes led us to make an erroneous assumption with regard to some of the reaction products and pathways.[190] Namely, when 2-bromo-7-methoxytropone (**87**) and **179** were refluxed in acetic acid, we obtained a mixture of **182b** (X = Br) (70%) and an orange-yellow compound B (5%), in addition to a minute amount of a dark-violet pigment, A.[191] The major product, **182b**, readily gave the ring-closed 6-bromo compound **187** when heated in acetic acid containing a trace of concentrated H_2SO_4. Then, similar treatment of the 3-bromo-2-methoxy compound **88** with **179** yielded, to our surprise, more than 12 colorful products, which were separated by HPLC (15 peaks) and TLC (12 spots).[192] We called these products **A**, **B**, . . . , **L** according to their decreasing R_f values. (Compounds **A** and **B**, either from **87** or **88** are identical). The structures of compounds **A–L** were determined by spectroscopic measurement and, in some cases, by X-ray analysis, as well (Chart VII).

Among these products, **J** and **G** are 1:1 condensation products (benzo[*b*][1,4]oxazinotropones). Other products (**A**, **B**, **C**, **D**, **F**, and **L**) are 1:2 condensation products and their secondary products formed by cyclization followed by dehydration or saponification (in the case of B).[191] These compounds can also be produced by the reaction of **179** and bromocyclohepta[*b*][1,4]benzoxazines (**187** or **188**), which is presumed to be a reactive intermediate in the previous reaction. Compound **L** was isolated as its HBr salt, which, upon basification accompanied by autoxidation, readily gave the chiral acetal **189**.[192,193] Because

187 **188**

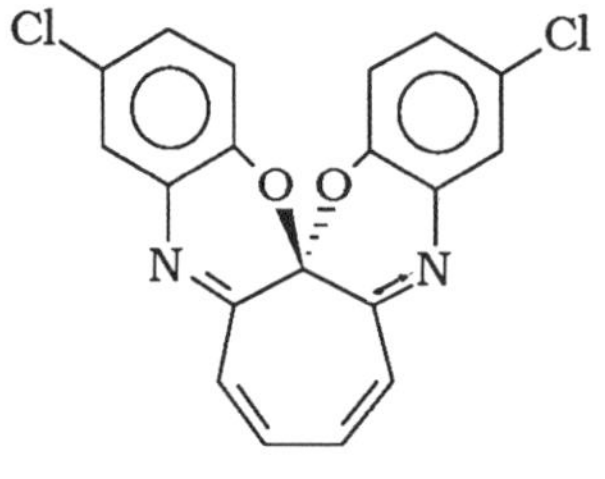

189a 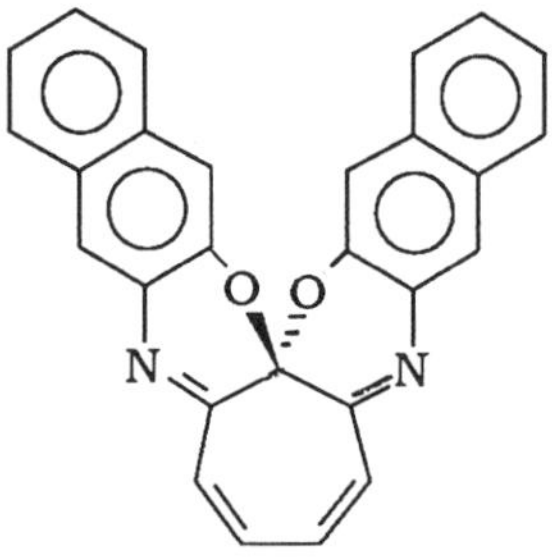 **189b**

the chiral acetal **189** (**F**) does not have an appropriate functional group, I asked Dr. Okamoto (Osaka University) to resolve it into optical isomers by using a chiral poly(triphenylmethyl methacrylate) [(+) PTrMA] column that he had developed. About 2 weeks after my request, the sample was returned to me with a letter of apology from Dr. Okamoto, saying that he had been unable to resolve it. He added that the two optical isomers were unusually close together, so much so that he could only observe one peak in the HPLC chart.

While I was reading this letter with disappointment, I received a phone call from Dr. Okamoto. He explained that to reconfirm his results, he had collected the substance from the large single peak and examined the $[\alpha]_D$s. To his surprise, he observed a very large optical rotation, $[\alpha]_D$ −4700, which was comparable to that of helicene. Although he could not locate a peak for the antipode, he finally found a small, flat peak at a point quite distant from the major peak. The $[\alpha]_D$ of the small peak was around +4600. Then, when a (−) PTrMA column was used, a sharp peak of the antipode ($[\alpha]_D$ +4700) was found.[194a] From the X-ray analysis[194a] of the (−)-3,12-dichloro derivative of compound **F** (**189a**), and also by theoretical calculations with CD (circular dichroism) spectra,[194b] we found that this substance (**189a**) and its naphtho analogue, **189b**), had an (*S*) absolute configuration. As for com-

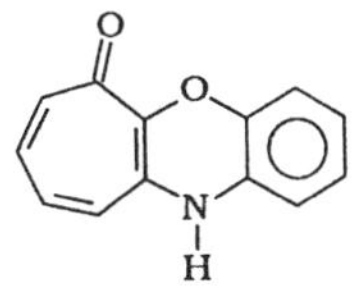

Chart VII

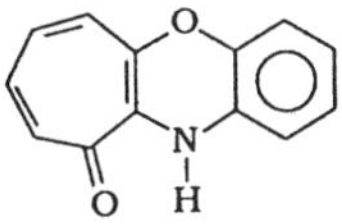

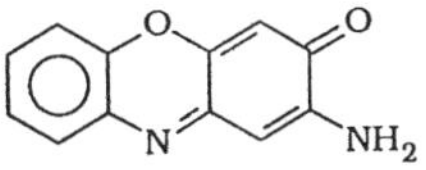

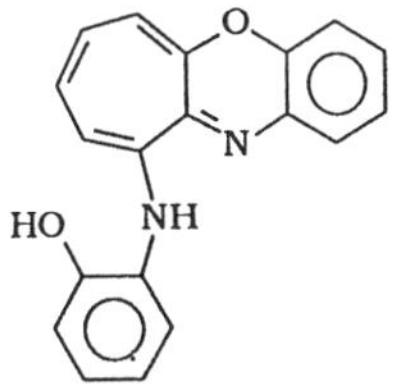

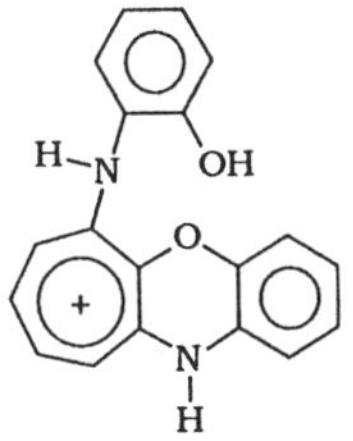

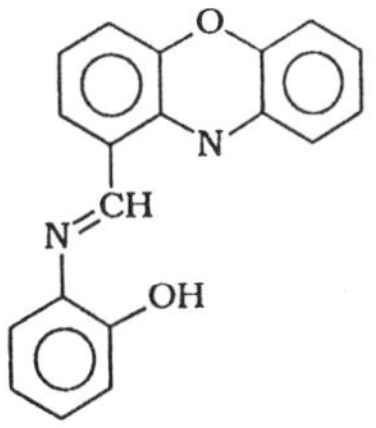

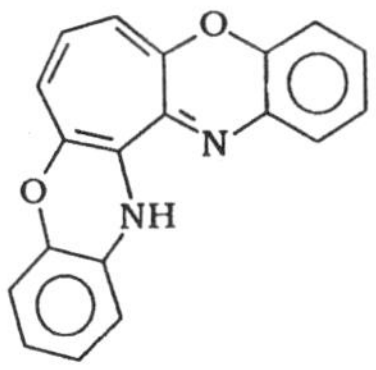

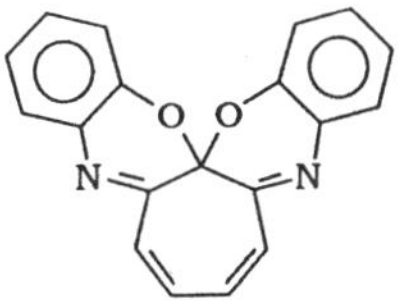

pound **D** (the precursor of **A**), we first assumed this to be a ciné substitution product at C-10 of **187**, but it has been proven later to be a normal substitution product of the 10-bromo compound **188**, which exists in equilibrium with **187** by the action of **179**[195] (*vide infra*). Compound **C** turned out to be a Schiff base produced by the attack of **179** at C-9 of **188**, which leads to a norcaradiene intermediate, and is easily saponified to give 1-formylphenoxazine, **190a** (**B**). Interestingly, the Schiff base of 4-formylphenoxazine (**190b**) instead is obtained as one of the main products when the reaction is conducted in the presence of a base (Dabco [diazabicyclo[2.2.2]octane]). H^1 is a coupling product of **179**.

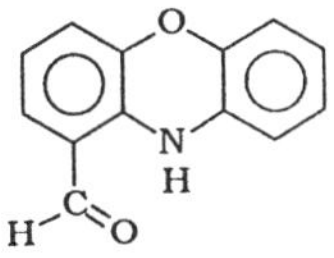

To find out the exact pathways of this highly competitive and unusual reaction, we then examined the condensation reaction of the 6-bromo compound **187** with 4-methyl-2-aminophenol (**191**). To our

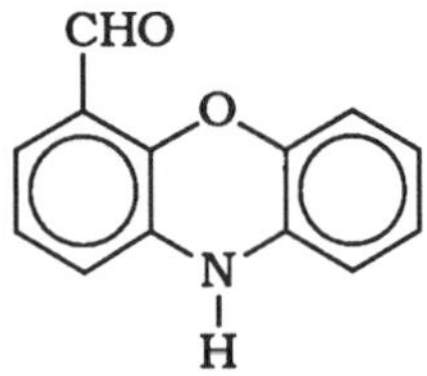

CHO

O

N

H

190b (CHO at C-4)

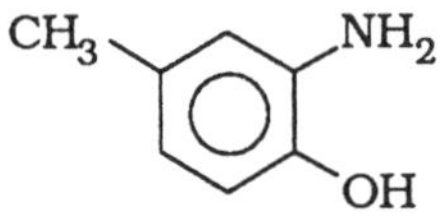

CH_3 NH_2

OH

191

surprise, we observed almost 30 peaks in the HPLC chromatogram. On careful examination, each of those products corresponding to the parent **L**, **F**, **A**, and **C** consisted of a set of a comparable amount of three compounds: the expected monomethyl derivatives, dimethyl derivatives, and parent compounds without any methyl group. This result suggested to us that prior to the substitution of the bromine atom in **187** by the amino group of **179** (or **191**), a certain kind of interconversion of the heterocyclic group was taking place that causes a complex and unprecedented intermolecular exchange of the entire heterocyclic system.[195,196a]

To avoid the complication caused by the bromine substitution, we first examined, with time-dependent HPLC, the reaction of two nonsymmetrically substituted isomers, **186a** and **186c**, with **179** in methanol at room temperature. A mixture of an equal proportion of both isomers is always produced, even if either of the single compounds **186a** or **186c** is used as the starting material. This finding can be explained in terms of the cyclic equilibrium involving the open-ring 2-aminotroponeimine intermediate **192**, which is formed by the nucleophilic attack of the amino group of **179** at C-5a of **186a** or **186c**.[195,196a] A theoretical calculation shows that C-5a is the most reactive site (followed by C-10) towards nucleophilic and radical attack in benzo[*b*]tropoxazine and its related ring system.[197a,b]

We then studied the reaction between the 6-bromo compound **187** and compound **179** (by the same method) in methanol or acetic acid. The ring-exchange reaction proceeds surprisingly fast in methanol, even at 5 °C. We isolated the 10-bromo compound **188** as pure crystals and determined its structure by NMR spectroscopy. From the examination of the reaction of **188** with **179** and other arylamines, it can be concluded that compounds **C** and **D** are obviously derived from **188** but not directly from **187**. If a strong base such as Dabco is added to a methanolic solution of **188** and **179**, the Schiff base of 4-formylphenoxazine, **190b**, is produced, because the exchange reaction in heterocyclic groups is known to be restricted under these basic conditions.[198]

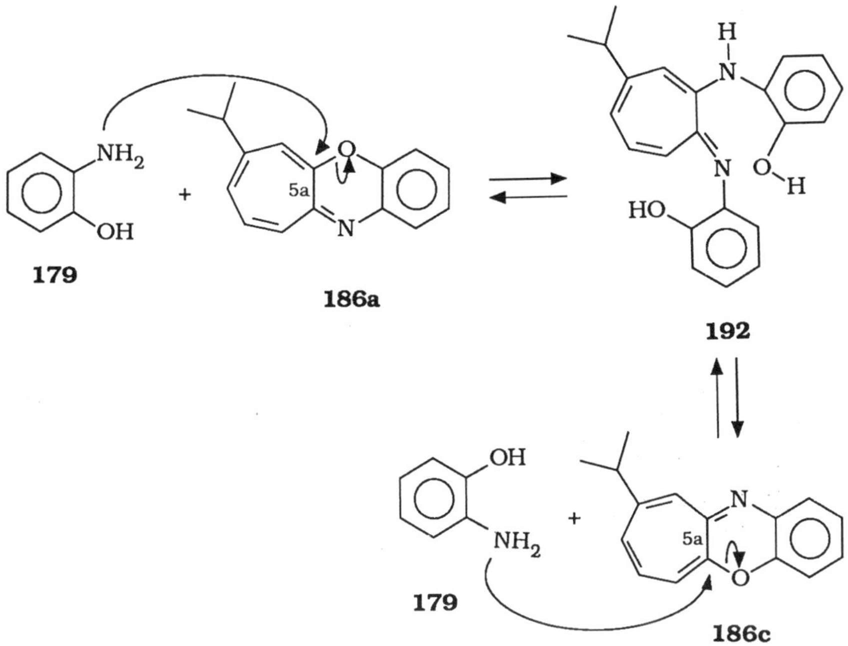

We found that the reaction of 5-bromo-2-methoxytropone, **89**, with **179**, especially under basic conditions, is far more complex than those of **87** and **88** with **179**. The structures of the main products are shown in Scheme XIV.[198] Particularly interesting reactions are the formation of **194** via the 8-bromo compound **193** and an intermediate a and of the *p*-tropoquinone derivative **196** from **195** (a free compound from the stable HBr salt) accompanied by an unprecedented intramolecular transposition of a heterocyclic ring on the seven-membered nucleus followed by dehydrogenations, as exemplified in Scheme XIV.[198]

Cyclohepta[*b*]benzothiazine and Its Analogues. Treatment of **183** with an excess of **180** in methanol at room temperature gives the thiazine analogue **177** in high yield.[196b] Similarly, both **183** and **177** can be led to the diazine analogue **176a** by the action of **181**. The reverse reactions from **176** to **177** or **183** do not occur, apparently because of the unfavorable nucleophilicity of the functional group. Compound **183** also reacts with aliphatic reagents, ethylenediamine (**197a**), 2-aminoethanethiol (**197b**), and 2-aminoethanol (**197c**) to produce, depending upon the

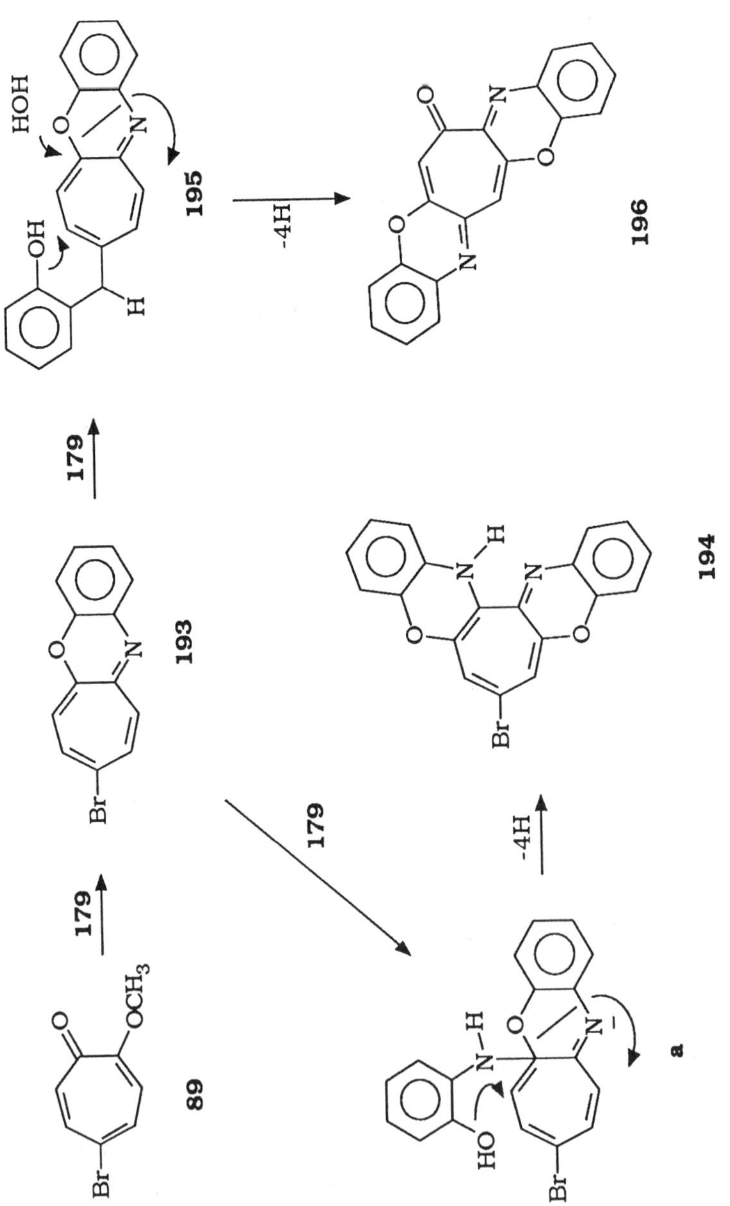

Scheme XIV

nucleophilicity of the functional group X (X = NH_2, SH, or OH), either the ring-closed product **199** (X = NH or S) or **198** (X = OH). On the other hand, **177** exclusively produces the rearranged product **200** on treatment with ethylenediamine, **197a** (Scheme XV).[196b]

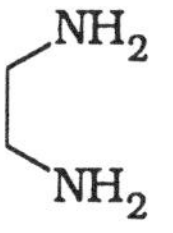

$NH_2CH_2CH_2NH_2$ **197a** $NH_2CH_2CH_2SH$ **197b** $NH_2CH_2CH_2OH$ **197c**

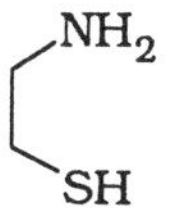

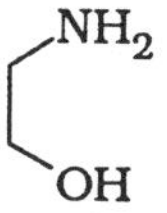

It has been astonishing to find that the S-substituted compound **201**, which is exclusively produced from **193** and **180** in methanol within 1 min at room temperature, gradually changes on standing in methanol to a mixture of compounds (more than 10 peaks by HPLC). These compounds consist of ring-closed products (at C-7 and C-9; **202a** and **202b**), dehydrogenated pigments (**203a** and **203b**) and their S-oxides, and 2- and 3-formylbenzoxazines (**205a** and **205b**) and their Schiff bases (**204a** and **204b**)[196b,197a] (Scheme XVI).

The heating of the ethanolic solution of **86a** (X = OCH_3) and **181** in a sealed tube at 80 and 120 °C gives *o*-aminoanilinotropone, **206**, and **176a**, respectively in high yields.[199b] Under acidic conditions, the dark-green salt of **178** is directly obtained. Compound **176a** is easily oxidized in air, especially under basic conditions, which produce dimeric compound(s). We obtained[183] white crystals by alumina chromatography (*vide supra*), which were in the form of an oxidized coupling product, as suspected by Fukunaga.[187] We could establish the structure of the main product as **207** by spectral data. Compound **207** is easily reduced back to **178** with zinc dust in acetic acid almost quantitatively[199b] (Scheme XVII).

We were also able to obtain *N*-monomethyl- (**208** and **209**) and *N,N'*-dimethyl (**210**) derivatives as green crystals. Not only the cations (**178**, **209**, and **210**), but also the free *N*-methyl compound (**208**) has almost the same green color and electronic spectrum.[197a,199b]

From NMR data and theoretical calculations,[197b,199b] we confirmed that these compounds are stabilized in the 6π-electron benzenoid–tropylium form (**178**, **208a**, **209**, and **210**), and their deep-green color is due to the intramolecular charge transfer and not the 16π-peripheral-electron system **178a**, as considered by Fukunaga.[187] We also found various heteroring transpositions and rearrangements in polyannulated cyclohepta[*b*][1,4]benzoxazines,[200a] as well as unprecedented facile exchange of the difunctionalized side chain in 2,3-dihydrocyclohepta[*b*][1,4]thiazines.[200b] It is very interesting that cyclohepta[*b*]franones and cyclohepta[*b*][1,4]benzoxazines and their S

Scheme XV

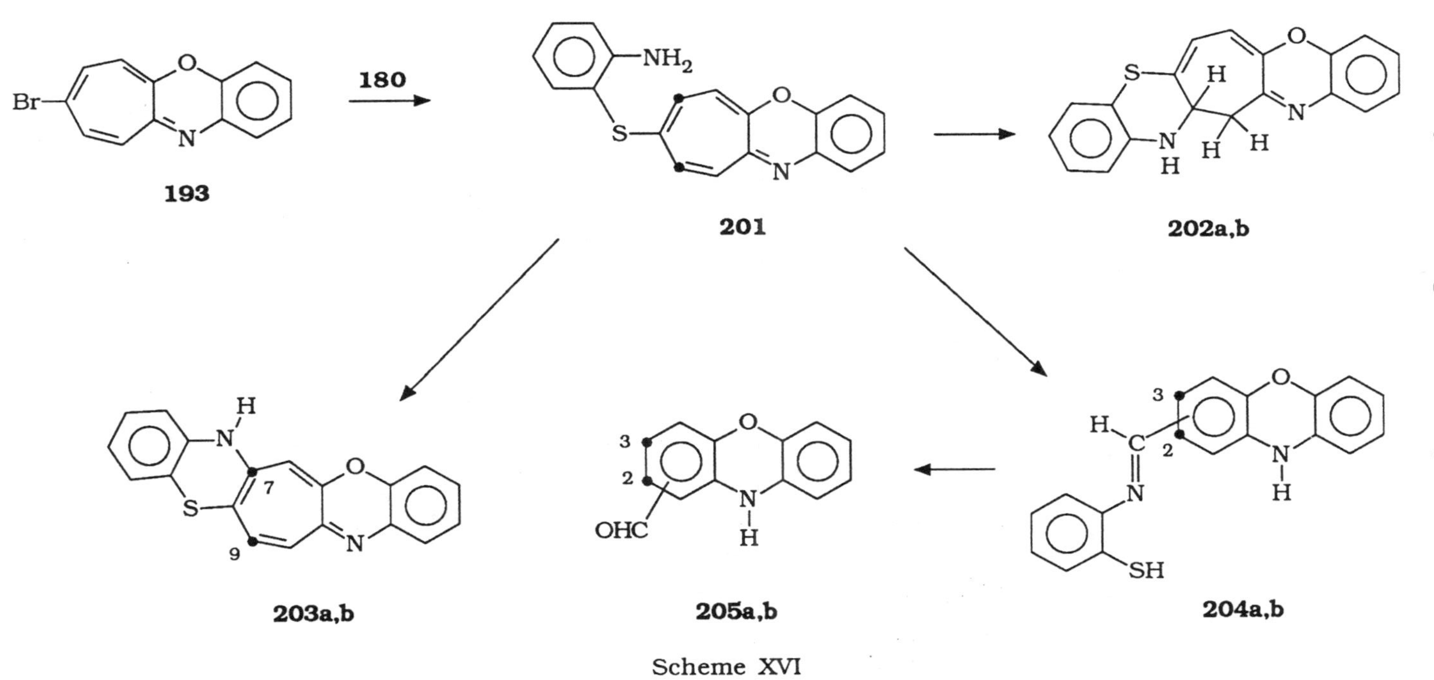

Scheme XVI

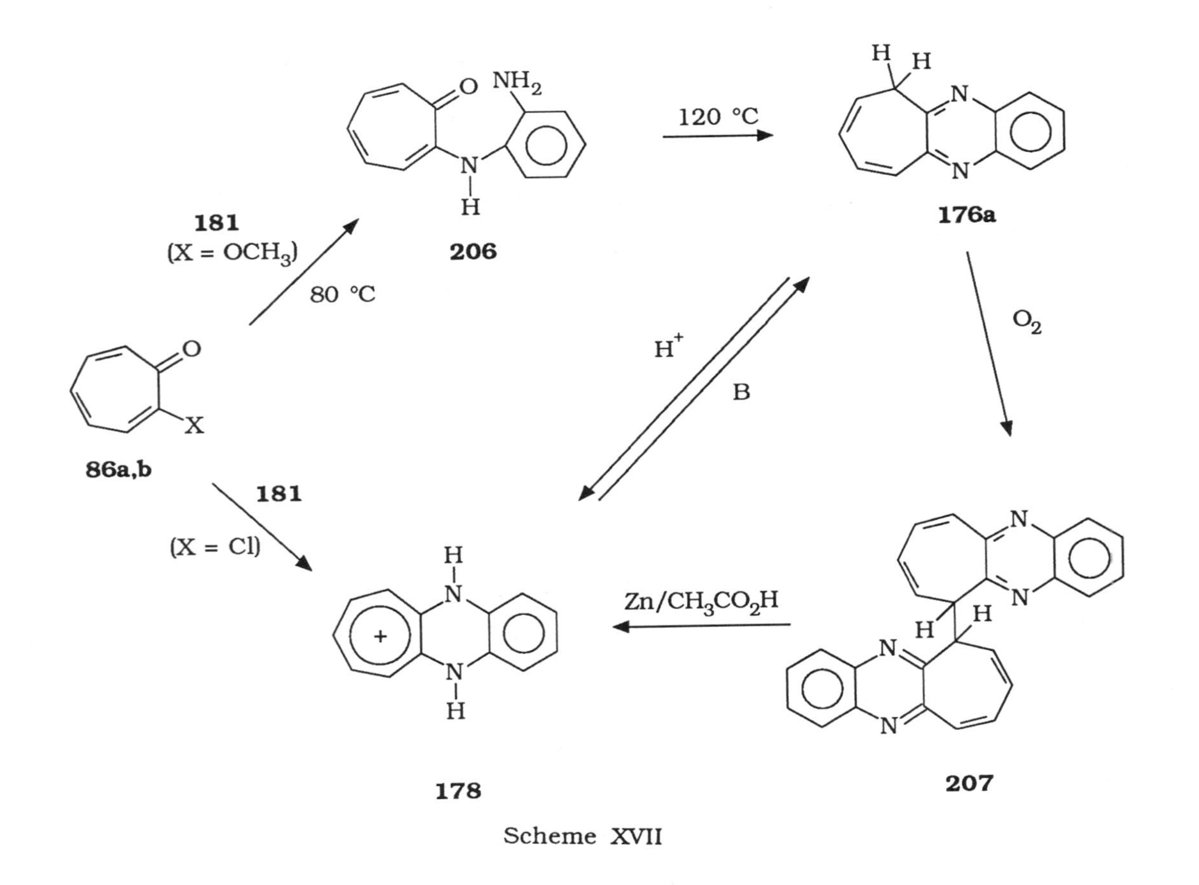
206
120 °C
176a
181
(X = OCH3)
80 °C
O2
H+
B
86a,b
181
(X = Cl)
Zn/CH3CO2H
178
207

Scheme XVII

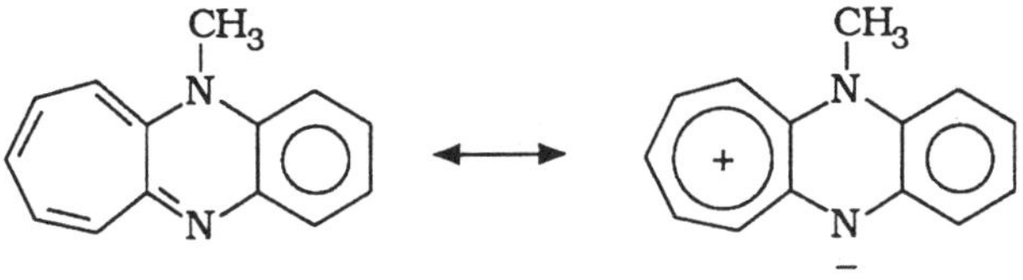

208 **208a**

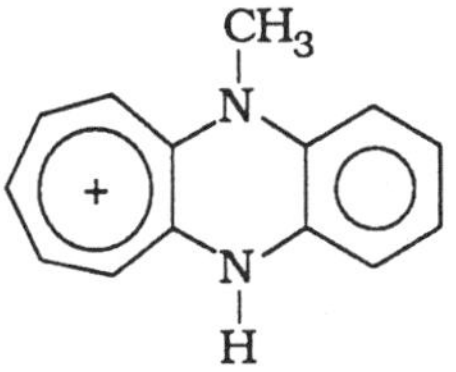

209 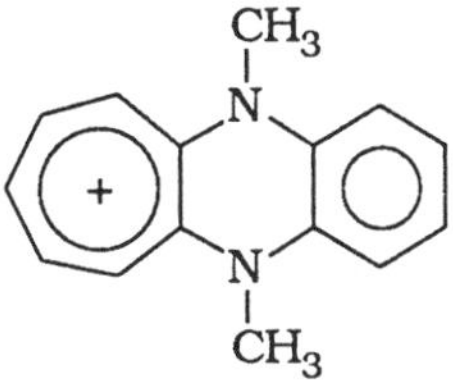 **210**

and N analogues show such unprecedented characteristics.[197a] Finally, we discovered the almost quantitative one-pot synthesis of various tropocoronands (*see* p 177) utilizing heterocycle exchange reactions of **183** with polymethylenediamines.[200c]

More recently, I have felt the necessity of introducing the method of electroorganic chemistry into the troponoid field. Various interesting results are now being obtained by Professor T. Shono and his colleagues at Kyoto University.[201]

Academic Activity after Retirement from Tohoku University

As I had accepted the regional editorship for *Tetrahedron* after retirement, I departed for a 4-month trip to Europe. The purpose of the trip was to gather information on the status of nonbenzenoid aromatic chemistry in Europe and the United States and to visit various friends in these countries. As usual, I was overwhelmed by the hospitality of my hosts. After retirement, I did not keep up with the chemical literature other than in my special field. A concerted effort was therefore made to attend and present 7- or 15-minute talks at annual meetings of the Chemical Society and to attend the various International Meetings that were being held increasingly often in Japan. I attended meetings full days to become acquainted with the forefront of organic chemistry; these full-day efforts, instead of tiring me out, were beneficial to my health. An unexpected bonus was that I made new friendships with the younger generation. I also renewed friendships with many old friends.

ISNA-2 held in Lindau, West Germany, in 1974 was a great success. I was asked to give a talk at the opening ceremony as the founding father of ISNA, the first symposium of which was held in Sendai. Erich Hückel, honorary president of the Lindau meeting, could not

13. 11. 1974

My dear collegue Nozoe,

Prof. Hafner sent me the report of the speeches, given on the II. ISNA-symposium held at Lindau. I am glad to hear how successfully and harmonically this symposium has succeeded. I believe that the further studying on nonbenzoid aromatic compounds will be not only interesting but also very fruitful for the different aspects of the chemical science.

I am very grateful for your merits in furthering all these advances.

I would be glad to here from you wether your opinion concerning this matter goes conform with mine.

With best greetings I am

your faithfully

E. Hückel

An excerpt from the comments of Professor Erich Hückel of the University of Marburg, written in 1974. Hückel, honorary president of ISNA-2, was pleased with the success of that symposium and interested in Nozoe's perceptions of the meeting.

The "Panda Effect", first discovered in Cupric Tropolone in 1951 (J. 1951, 1222)

With best regards.

J. Monteath Robertson

Nozoe visited Professor Robertson in Glasgow in 1966. The "panda effect" refers to the resemblance of the electron density map of cupric tropolonate to a panda.

attend because of poor health, but having heard of the success of the meeting from Professor Hafner, the chairman, he wrote me expressing his gratitude for my contribution in promoting the molecular orbital theory. Hückel wrote on the letterhead of ISNA-1, which he had gotten from Hafner; he was probably pleased with it because the ISNA-1 symbol was modeled after the $(4n+2)\pi$ Hückel rule. I am sorry I missed the opportunity to see him. The word "nonbenzenoid" was replaced by "novel" during ISNA-3, held in 1978 in San Francisco, with Professor Breslow as the chairman.

ISNA-4 was organized in Jerusalem by Professor Israel Agranat, where I delivered one of the plenary lectures. During the excursion, we climbed from the world's lowest restaurant, located 300 m below sea level to the top of Massada, with a temperature of 40–45 °C; the heat did not overly bother me because it was dry, but I became famous for being the first to make it to the Massada summit. In the talk at the symposium banquet, Professor David Ginsburg mentioned this Massada event, but referring to the fact that many of my recent studies were performed at the Kao Soap Co., also said that "Nozoe has cleaned up [with soap] a most interesting area of tropolone chemistry." After the meeting

Homage to Professor T. Nozoe, one of the fathers of non-benzenoid aromaticity:

two groups of ^{1}H-NMR signals
external $\delta = 6.7 - 8.2$ (14H)
internal $\delta = 4.5 - 4.9$ (6H)

benzo[18]annulene: non-benzenoid diatropicity!

Heidelberg, 27.4.79 Heinz A. Staab

Nozoe was a guest in the home of Professor Staab in Heidelberg, 1966, and in 1979 at the university.

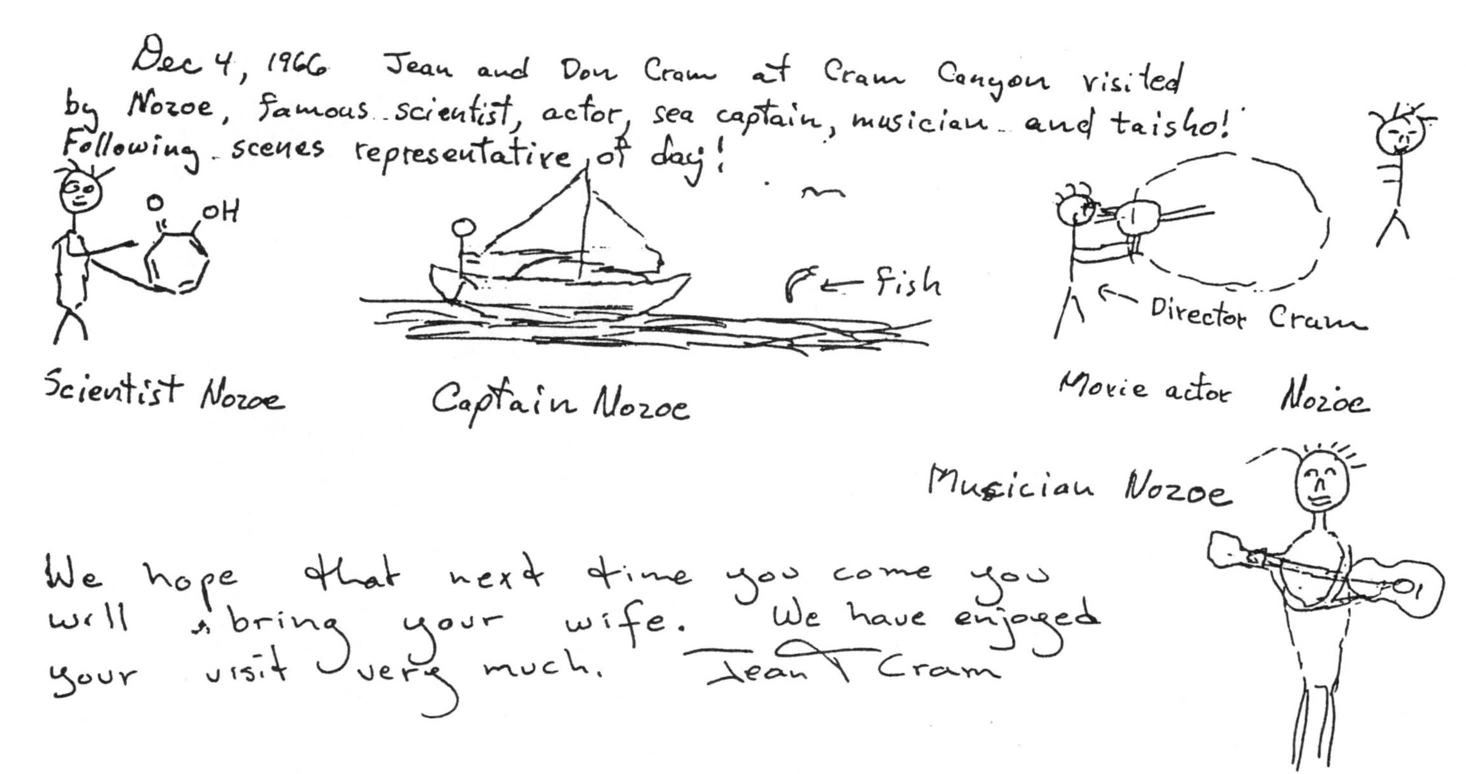

Nozoe visited Professor Cram (Nobel Prize, 1987) and Mrs. Cram in their Pasadena, California, home, Cram Canyon, December 1966. The illustrations refer to Nozoe's interest in sailing and Cram's in music. Cram was enjoying the movie camera that he received as an early Christmas present from his wife.

R.B.W./1927

cobyrinic acid abcdeg hexamethyl ester f-nitrile
beautifully crystalline sample of totally synthetic material obtained during Professor Nozoe's visit—and at the time of 90% eclipse of the sun!
10. July 1972

森
王

RBWoodward

Nozoe visited R. B. Woodward at Harvard University in 1972. The structural formula (top right) was assigned for benzene by Woodward in 1927, when he was only 10 years old. His signature in Chinese characters, taught to him by Shô Itô, reads "king of wood."

With Louis Fieser on his 65th birthday, April 7, 1964, are Mary Fieser, Tetsuo Nozoe, and Professor Shiokawa, chairman of the chemistry department at Tohoku University in Sendai.

So many years since first we met in Katonah in 1953. And now my dear, old and so very young friend, in Tokyo! This first day and first dinner in Japan, I shall never forget, nor can I thank enough.

The early thoughts of rapid equilibrium in

have grown into the arcane intricacies of the not-obviously-concerted rearrangements and reactions, of which the most recent begs for understanding:

0.48 0.41 1.00 1.00 0.60 0.22

In affection and ever-growing admiration, dear Nozoe.

William von Eggers Doering
3 September, 1977.

William von Eggers Doering signed the autograph book on his first visit to Japan in Tokyo, September 1977, on the occasion of the 26th IUPAC Congress.

It was a stimulating, exciting, challenging, and rewarding venture. I am extremely pleased to be able to see a few new research areas have been opened through this program.

Y. Kishi

A new project was initiated and hopefully will lead to something new.

Thiolase 4 × 41K

CoASH

NADPH

AcAc CoA reductase 4 × 25K

NADP

PHB Synthase

$n \simeq 10{,}000$ PHB

June 1, 1988 at Kyoto

S. Masamune

Professors Kishi of Harvard and Masamune of MIT, two Japanese plenary lecturers, signed at the 16th International Symposium on the Chemistry of Natural Products in Kyoto, June 1988.

May 26, 1988
25* years of participation in this wonderful travel book! The pleasure to find, in the preceding pages, so many signatures of common friends. The pleasure of meeting again Prof. Nozoe on his home ground. The pleasure of coming again to an important IUPAC meeting in Kyoto. The pleasure to be here, with old and new friends!

Guy Ourisson

* Alas! 1988 − 1953 = 35 years, not 25.... 35 years since you first came to Paris... Yesterday!

OH Who will find it? where? when?

Professor Ourisson visited Tokyo in May 1988, to lecture at the Takasago Perfumery Company, Tokyo Research Institute.

To Prof Nozoe pioneering scientist, most friendly and gracious host, and an inspiration to us all – with warmest thanks and all best wishes.

OHC CHO H H OH

E. J. Corey
30 October, 1965
Matsushima

with Shigeo Nozoe
1963–1964

Excursion to Matsushima, near Sendai, in October 1965. From left: H. Uda, Tetsuo Nozoe, E. J. Corey (Nobel Prize, 1990), and Shô itô. (Photograph courtesy of K. Nakanishi.)

JW Cornforth 23/1/81
Happy return!

OMe
OH
OH
Me
OMe
MeO
O
O
P
CH_2P O O$^-$ OH
MeO
$Me_3\overset{+}{N}CH_2CH_2O$
$OCH_2CH_2\overset{+}{N}Me_3$

The hopeful catalyst.

Nozoe makes a haiku:
Azulene from tropolone,
Seven and ten!

For azulene-making from tropolone
The cash was provided by soap alone

Me_3Si, $SiMe_3$, Me_3Si, $SiPh_2I$ C + Ag^+ → Me_3Si, $SiMe_2$, Me, Me_3Si, $SiPh_2$ C +

MeOH ↙ ↘ NO_3^-

Me_3Si, $SiMe_2OMe$, Me_3Si, $SiPh_2Me$ C Me_3Si, $SiMe_2ONO_2$, Me_3Si, $SiPh_2Me$ C

Colin Eaborn
Sussex 23 January 1981

Tetsuo :

Welcome again to U.K. and Brighton!

After 30 years I have returned to tropolones with a lot of nostalgia – see Alan Morgan's entry on next page. I hope your very strenuous European trip is successful and you can soon return to Japan for a rest! We are still very busy studying the mode of action of the B_{12} enzyme systems :

Looking forward to arranging your 4th U.K. trip

Alan Johnson 23.1.81

(from A. nidulans)

Alan Morgan Sussex 23/1/81

Nozoe's unusual lecture to an audience of five, including three eminent chemists, took place at the University of Sussex in Brighton, January 1981. The signatures on p 198 are those of Sir John Cornforth (who shared a Nobel Prize with V. Prelog in 1975) and Professor Eaborn. At Nozoe's request, Sir John wrote a haiku. On this page are the autographs of Professor Alan Johnson and Alan Morgan, a graduate student who took care of the slides for the two-hour lecture.

Exploring Israel during ISNA-4, August 1981. Top: Tetsuo and Kyoko Nozoe at the top of Mt. Massada, which stands only 40 meters above sea level but which rises nearly 400 meters from its base at the Dead Sea. Bottom: Nozoe and Professor Murata of Osaka University pause before the ruins of an old fortress on the mountain.

My interests are mass spec., n.m.r,
The power of these stretches far,
The choice for me was not easy,
But my syntheses weren't so breezy,
Leading to gums or crystals doughy,
- Not at all like those Nozoe.

If I ever return to the fold,
Trying syntheses new, crazy, or old,
My azulenes won't rise to par
And my tropolones will come out as tar.

To the who worked in Taiwan,
And later impressed all Japan,
With his crystals blue, white and black,
- We say to you please come back.

24th Sept., 1966.

Dudley Williams
Pat. Williams
Coach & Horses
Trumpington.

Ostreogrycin A.

Yes, again!!
F. Sondheimer

Not to mention
B. Sondheimer

Fortunately Dr. Williams is better
at chemistry than he is at poetry!
His formula is good but his rhymes are shocking.
To be with Nozoe is always a pleasure — may we be together
again soon. Todd 25/9/66.

Dr. Williams contributed a structure and a poem, Professor and Mrs. Sondheimer concurred with the sentiments expressed, and Lord Todd commented upon it all in Cambridge, 1966.

ISNA 4.
Jerusalem, August 30 - September 4, 1981

To Professor Tetsuo Nozoe
A Founding Father of ISNA and the
Doyen of Novel Aromatic Chemists
with great admiration

Mordecai Rabinovitz Israel Agranat
Ilana Agranat

At the ISNA-4 Farewell Dinner I wished, in the name of all of the participants that when we all are 79 years young we should all be as sharp, intelligent and enthusiastic as you are, Prof. Nozoe, as you have demonstrated to us in two ways: 1) By "cleaning up" (with soap) a most interesting area of troponoid chemistry 2) By being the most nimble climber up Massada during the Symposium tour.

Now that we are enjoying your presence in our house in Haifa, and after you have now met our immediate family, you and Mrs. Nozoe are now honorary members of our family. We wish you good health and continued happiness and we hope to meet you soon again.

לשנה הבאה בחיפה
(Next year in Haifa)

David Ginsburg

Reminiscences from ISNA-4 in Jerusalem, September 1981: Professors Agranat (chairman), Rabinovitz (general secretary), and Professor D. Ginsburg, with whom Nozoe stayed while in Haifa.

10π aromatics
— just organic!

almost black!
a 14π azulene

S_4N_4 + 2 Ph–C≡C–H $\xrightarrow[\text{Toluene}]{\Delta}$

how?

purple.
v. stable
soln!

Wonderful to see you
again — getting younger all the time!
Charles Rees, Imperial
College London
21 July 1987, Kobe

Professor C. Rees of Imperial College in London signed at the First International Conference of Heteroatom Chemistry in Kobe, July 1987.

I was invited to his home in Haifa for his birthday (it is a pity that he passed away in 1988 before we were able to meet again). Ginsburg received the Hoffmann Prize immediately after me.

ISNA-5 was held in St. Andrews, Scotland, with Professor Rees as chairman, where I again met with acquaintances stemming back from ISNA-1. The meeting ended on Dr. Lloyd's 65th birthday. He received many visitors, including his friends Professors E. Vogel, H. Volz, P. Garatt, and several participants from Japan. Professor Rees gave a lecture on a 10π aromatic system containing only a single carbon atom; subsequently, he expanded this to a 14π system at the First International Conference on Heteroatom Chemistry held in Kobe, Japan, in 1987. ISNA-6 returned to Osaka in 1989, and with Professor Murata as head, was a great success. In addition to old-timers, we saw an expansion in the field with expert participants from inorganic chemistry and quantum mechanics. Professor B. Gimarc presented a lecture on carbon-less inorganic aromatic compounds and on porphyrin analogs. ISNA-7,

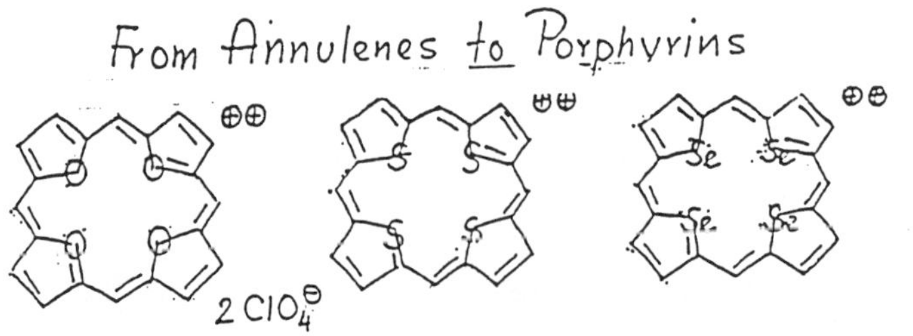

Dear Professor Nozoe: It is with great pleasure that I recall ISNA-1 in Sendai in 1970 held in your honour after your retirement. It is an even greater pleasure to meet you again, in fine spirits and good health, at ISNA-6 in Osaka almost twenty years later.

August 1989

Emanuel Vogel

C_{60} Buckminsterfullerene

First thought of in Japan by Osawa + Yoshida

Very Best Wishes

Harry Kroto

University of Sussex
Brighton
England.

Harry Kroto – with very best wishes
We have very much enjoyed Japan.

Aromaticity is an important concept for all of chemistry, including inorganic examples as well!

Ben Gimarc
University of South Carolina

Plenary and invited lecturers at ISNA-6 in Osaka in August 1989: Professors Vogel, Kroto, and Gimarc.

which I would like to attend, will be held in Victoria, Canada, in 1992. We shall see.

I visited the United Kingdom for 2 weeks in January 1981 in an exchange program between the Japan Academy and the Royal Society of Chemistry, where I was introduced to the most recent studies being carried out at various institutes; I gave six talks. I had an interesting experience at the University of Sussex, Brighton, when I visited Professor Alan Johnson, who at that time was delighted with discovering a new natural tropolone. As it was my first stop on this trip to the United Kingdom, no public lecture was planned because he thought I might be tired. However, he asked whether I minded giving an informal talk in his office to several faculty members. When I asked which of the two prepared talks, he said both; I ended up giving two lectures for 2 hours in his small office to Professors Cornforth (Nobel laureate), Johnson, and five other staff members. It was a unique experience. When I spoke in writing with Cornforth, who had difficulty in hearing and speaking, he presented me with the limerick shown on p 198. I traveled alone to the United Kingdom this time because I feared the cold temperature, but it was like spring.

In late March 1988, I visited Sweden for 2 weeks as an exchange scientist between the Japan Academy and the Royal Swedish Academy of Sciences. Because it was very cold the previous 2 years, I again traveled alone. The General Secretary, Professor Ganelius, took great care of me. I was disappointed that I was not able to see Professor Erdtman, an old friend whom I had looked forward to meeting, but was glad to be with Mrs. Aulin-Erdtman at Professor Lindberg's dinner party at his home.

I attended the annual meeting of the Royal Swedish Academy with another foreign member; King Carl Gustav, XVI, entered the small

The senior members of the Royal Swedish Academy of Science wait for the entrance of King Carl Gustav, XVI, March 29, 1988.

conference room where about 20 of us were waiting; after shaking hands with everyone he entered the hall. The ceremony of the annual meeting was quite simple. Only three people were on stage: the academy president, the general secretary, and an executive member.

Visiting Stockholm, Sweden, in April 1988. Top: Professor Ganelius, general secretary of the Royal Swedish Academy of Science, and Mrs. Ganelius. Both of them were extremely kind and saw to Nozoe's needs during his stay in Stockholm. Bottom: Dinner with Professor Lindberg of the University of Stockholm. He is flanked by Mrs. Gunhild Aulin-Erdtman and Mrs. Lindberg.

The banquet was held in a restaurant because of the large number of participants. Of about 20 tables, I was assigned to Table B, where I sat across from the queen, who was flanked by the academy president and the head of the Supreme Court; the king was seated at Table A. I could only guess (because all the speeches were in Swedish), but judging from the laughs, all talks, including that of the king, must have been very

Kai Siegbahn March, 25, 1988
Last time we met was in Okazaki
I had been told about the "7-ring" and
was excited to meet him personally!
This time we met with the Japanese
Ambassador and Ambassadress few days
before going to Japan to revisit old
friends there!
Anne-Britt Siegbahn

At the Japanese embassy in Stockholm, March 1988, during a visit to Sweden on an exchange program between the Japan Academy and the Royal Swedish Academy of Science. From left: Professor Siegbahn (Nobel Prize in physics, 1924), Ambassador Y. Nomura, and Tetsuo Nozoe. The signature in the autograph book was that of Professor Siegbahn.

At a general meeting of the Royal Swedish Academy of Science in Stockholm in March 1988. From left: Professor S. Gronowitz of the University of Lund; Professor F. Lindqvist, president of the Academy; and Tetsuo Nozoe.

At an ISNA-6 banquet in Osaka in August 1989. From left: Tetsuo Nozoe, Mrs. Vogel, Kyoko Nozoe, Professor E. Vogel of the University of Cologne, Mrs. and Professor I. Murata (chairman, Osaka University).

humorous. There were toasts after each talk. The entire atmosphere, like that of a party in Taiwan, was very congenial. We had coffee in a small room, where I was again seated with the king and queen, the academy president, Professor Ganelius, and five others. I took a picture of the king and queen after receiving permission from the queen. However, I refrained from conversation because of my trouble in hearing. The royal couple left about 11 p.m. after shaking hands with all participants: quite different from the Japan Academy.

Special celebrations are held in Japan to honor one's 88th birthday. In October 1989, 90 former students from Taihoku Imperial University (in Taiwan) and Tohoku University held a celebration party for me. At the end of 1989, I was scheduled to visit Taiwan as Tamkang Chair Lecturer; here again 65 ex-students from the Taihoku Imperial University period to the present National Taiwan University planned a get-together for me. An international conference was held in May 1990 in Sendai for my 88th birthday. Professors Breslow, Hafner, and Vogel kindly attended as plenary lecturers, while Professors Nakanishi, Masamune, Mukaiyama, Sakurai, and Murata consented to present invited lecturers. I am indeed grateful for all this. As long as I am healthy, I intend to do my best to contribute to the advancement of science, and in this manner, to repay the kindness and assistance extended to me by my colleagues over the years.

Conclusion

Reflections on My Work

During my past 65–70 years in organic chemistry, my scientific achievements have been varied and have continually been modernized with the appearance of modern analytical equipment (especially HPLC) and recording spectrophotometers (UV, IR, and NMR). In the earlier years, I studied saponins, sapogenins, and wool wax. In those days, we considered ourselves fortunate to isolate pure materials and to be able to identify some of them. Perhaps our most significant findings in these areas were the planar structures **12a** and **12b** that we proposed in 1937 for oleanolic acid and hederagenin, the most widely distributed components of sapogenin. This study led to the revision of the previous formulae. Our structures have been subsequently proven accurate and are still in use. In 1939, we presented the structures of two series of branched-chain fatty acids (**17** and **18**) in wool wax in short communications, and in 1941, we presented short abstracts on the partial structures (**21a** and **21b**, respectively) of α- and β-lanosterol for the first time.

In 1936, I isolated hinokitiol from the essential oil of the conifer, and in 1943–1947, I reported the seven-membered-ring α-trienolone structure independently from Professor Erdtman. On the evidence of chemical reactivities, I considered this seven-membered ring to be a new type of aromatic compound. These findings triggered off the emergence of troponoid chemistry, which eventually opened up a wide area of the study of nonbenzenoid aromatic chemistry. After 1949 we synthesized hinokitiol and its parent compound tropolone and systematically studied the chemical reactivities, which included the very characteristic electrophilic substitution, addition, and rearrangement reactions. Moreover, we found that the amino derivatives of tropolone obtained by the reduction of the corresponding nitro or azo compounds undergo Sandmeyer-type reactions, whereas in contrast to the benzenoid system, the Friedel–Crafts-type substitution does not take place.

In 1951, independently and simultaneously with overseas scientists, we also succeeded in synthesizing tropones. These compounds have more olefinic character than the corresponding tropolones and tend to give rise to addition–elimination reaction products. From these tropones, we then established a convenient preparative route to tropolones through 2-aminotropones. Starting from the so-called "reactive troponoids" (**86**), which bear a good leaving group at C-2, we established versatile preparative routes for a wide variety of interesting troponoid compounds by using normal and ciné substitution reactions and heterocyclic-ring formation.

Subsequently, we found (independently from overseas chemists) that tropylium ions and tropyl ethers readily react with nucleophiles to give 7-tropyl derivatives, which can lead to various troponoids via thermally isomerized tropyl compounds. Starting from tropylium ions and tropones, we also synthesized a novel type of thio- and iminotropones, heptafulvenes, and cross-conjugated quinarenes. With these syntheses, we contributed considerably to the expansion in this area. Since 1948, with the cooperation of physical chemists in Japan, we have been helping to clarify the physical properties of the troponoid nucleus to a considerable extent.

Since 1955, we have been developing novel and convenient one-pot polysubstituted-azulene syntheses starting from troponoids via cycloheptafuranones (**122**). In relatively recent years, we have intensively studied the autoxidation and peracid oxidation of guaiazulene and other azulenes and established the structures of a large number of various types of oxidation products. Remarkably interesting results obtained by our recent study were the discovery of unprecedented intermolecular one-carbon transfer during oxidation of azulenes, as well as intermolecular exchange and intramolecular transposition of heterocyclic groups in cyclohepta[*b*][1,4]benzoxazine and its related compounds, much to our happiness and amazement.

Various types of troponoids and azulenoids have thus become readily accessible. In particular, these troponoids have been found to possess concurrently both aromatic and olefinic characteristics and thus exhibit extreme diversity in their chemical reactivities. These novel nonbenzenoid aromatic compounds differ exceedingly from the benzenoid aromatic compounds.

As I look back over the past 60 years of my research, the most striking change has been the rapid advancement of analytical instrumentation and the accompanying changes in experimental procedures. These phenomena have been especially noticeable in the past 40 years since the end of the war. When I began my structural elucidation studies of natural products, the separation and analyses of similar compounds, such as isomers, required a large quantity of materials and took a long time. Except for the separations based on acidities, the most common methods used for the separations were fractional distillation (which is fairly inefficient) and recrystallization as functional derivatives. Neutral polyhydroxytriterpenoids, the first subjects of my research, contained isomers and similar compounds with various numbers of hydroxyl groups in the same skeleton. Thus, I was able to separate them by repeated fractional recrystallization after converting the initial substances into acetonides and their esters.

At that time, although alumina and silica gel column chromatography were available, these absorbents were very expensive. Thus, it was thought impractical to use these columns for the large-scale separation of the components of wool waxes. Fortunately, when I visited an aluminum factory in Kaohsiung (Taiwan), I received 20 kg of pure alumina as a gift. I was able to activate this alumina and use it for large-scale column chromatography. However, the solvents required for the chromatography, diethyl ether and petroleum ether, were very difficult to import to Taiwan from Japan. Fortunately again, I was able to obtain petroleum ether (bp 60–80 °C) from a natural gas well at Kinshiu (Formosa). Using these materials, I was able to separate completely large amounts of wool wax acid ester into three groups: one that contained hydroxyl groups and two that did not.

The acidic portion of the essential oil of the heartwood of the *taiwanhinoki* contains such substances as *l*-rhodinic acid ($C_{10}H_{18}O_2$), hinokitiol ($C_{10}H_{12}O_2$), and carvacrol ($C_{10}H_{14}O_2$), whose boiling points and acidities are very much alike. Separations using the conventional procedures were not easy. Nonetheless, I was able to isolate a small amount of hinokitiol from the acid mixture by using the metal-complexing properties of hinokitiol. At that time, the structural study of an unknown compound was based on derivatization to known substances by various chemical reactions. For example, the determination of the mother skeleton of a polyterpenoid or steroid was carried out by the determination of the structure of an aromatic hydrocarbon that is produced through dehydrogenation by heating the substance at a high temperature (250–300 °C) with S or Se. However, when the substances were treated under those drastic conditions, the mother skeletons were often rearranged. Therefore, definite structural elucidation had to be based on various oxidation and reduction methods. In those days in the 1930s, time-consuming measurements of UV spectra were not used for the structural determination of colorless (or pale-yellowish) compounds. However, I used this technique often and was able to correct errors in the already published structures of oleanolic acid and hederagenin.

The molecular formula of hinokitiol was very simple, yet its structure and properties were quite different from that of any previously known substance. Despite its simplicity and because it underwent addition–elimination reactions to give substitution or rearrangement products with electrophilic reagents, the determination of its structure was very difficult. Because all of the derivatives had novel skeletons, it was almost impossible to directly correlate them to the known substances. Nonetheless, the structure of the amino compound, which was produced by the reduction of nitro or arylazo derivatives, was correlated, fortunately, to that of other substances, because these amino compounds, like benzenoid compounds, were susceptible to the Sandmeyer reaction. Also, I was able to determine the structures of various substitution products by using the facile rearrangement of dinitrohinokitiol to benzenoid compounds and by measuring dipole moments.

When I returned to Japan, IR and NMR spectroscopic techniques were not yet available in Japan. Therefore, I had to use magnetic susceptibilities in the study of the aromaticity of troponoids, in addition to chemical reactions. When in early 1954 the first IR spectrophotometer was installed at the University of Tokyo Analytical Center, I was shocked to know that this technique allowed one to rapidly check the presence of various functionalities, including conjugated nitriles and carbonyls. Moreover, when our laboratory received an NMR instrument around 1960, I was shocked again that the number of hydrogen atoms and their connectivities could be clarified. The approach and concept of structure determination thus was to change dramatically; microquantities could be handled, and the entire methodology of structure determination was drastically altered.

The troponoid adducts (with reagent, catalyst, and solvent) we had been studying could not be analyzed by gas chromatography because of their thermal lability. The introduction of HPLC, therefore, greatly facilitated the analyses of various compounds coexisting in the reaction mixture. Reminded of my experience 55 years ago, when I visually followed the color reactions of monosaccharides with a manual photometer, I found that following the reactions of troponoids and azulenoids at suitable time aliquots disclosed the occurrence of many unexpected phenomena in solution. Namely, it was disclosed that, from the beginning, many reversible and irreversible competitive reactions were taking place among the substrate, reagent, catalyst, and solvent (such as alcohol). It was possible not only to measure the retention times and electronic spectra of more than one compound contained in each peak but also to isolate and, in many cases, to characterize them by means of MS and NMR techniques. It was impossible to imagine that such feats could be achieved in earlier days before and during the war. I was deeply impressed by these results.

My research topic is unrelated to the concerns of the Kao company, and moreover, the Kao Research Laboratory in Tokyo is not an organic chemical institute. Therefore, except for the HPLC and UV studies, all other spectroscopic measurements (IR, MS, and NMR spectra) had to be carried out at other places. In many cases, we had to wait for 3 to 4 weeks to get results. Furthermore, because the resolution of spectroscopic instruments was crude in the early days, unexpected products were obtained often; this situation led sometimes to the proposal of erroneous structures and mechanisms. Fortunately, the equipment at my collaborative laboratories has improved gradually, and measurements are carried out more rapidly. Recently, the research is carried out at several collaborative laboratories, and I am involved only in checking the results by time-dependent HPLC and in discussions of results.

Looking back on the results of my past research, particularly after my retirement from Tohoku University, I cannot help feeling that the results are exceedingly naive, judging from the highly advanced level of current theory and experimental techniques. This naiveté exists mainly because the research had to be done by staff members who had no experience in this field and undergraduate students. Nevertheless, I believe we could manage to find a series of interesting reactions.

I am convinced that with the modern instruments available these days, it would be possible for modern chemists (and not old timers like myself) to elucidate complex mechanistic paths and control reactions at will. From this point of view, I strongly feel that research in the field of troponoids and azulenoids provides excellent training for young chemists and students. Recently, several troponoids and azulenoids with interesting biological activities or physical properties were prepared. The compounds offer limitless synthetic possibilities. I am confident that the days will arrive when such compounds will contribute to the benefit of mankind, as I lectured at my first lecture abroad (in Tübingen in 1953). To pursue research in this field properly, it is essential not only to focus on the synthesis of a particular target compound but also to examine the numerous intermediates and competing reactions as a function of time by HPLC and other means. It is the right, obligation, and duty of chemists to show curiosity and affection toward all phenomena surrounding them.

Reflections on My Career

I have now passed that special 88th birthday. I become overwhelmed when I look back and think about all the things that have happened since I started my research in organic chemistry almost 70 years ago. My most important period as a chemist, from age 35 to 55 (1933–1958), was also the most trying period for Japan because of the immediate and long-term effects of the war. Furthermore, since then, organic chemistry has undergone a dramatic change in its content and methodology. The progress has been astounding.

My interest in chemistry has been triggered by the influence of my middle-school chemistry teacher Mr. Kiyoshi Hasegawa, the creative atmosphere at Tohoku University, and the influence of my two mentors, Professor Majima (Tohoku University) and Dr. Kafuku (Formosa). My mentors taught me the importance of simple curiosity, intellectual naiveté, and originality of thought. These factors taught me the importance of independent thinking and motivation.

The fact that I went to Formosa, after the strong advice of Professor Majima, without even going to graduate school had a fundamental influence on my life. If I had stayed at Tohoku University and continued my research there, it is quite possible that tropolone chemistry would not have reached its current stage of development. The research in polyterpenoids and wool wax in which I became totally absorbed during my early 30s did not fully mature and remained unknown to the rest of the world because of the war and because the short communications were written in Japanese. Fortunately, I had found hinokitiol as a minor constituent of *taiwanhinoki* wood oil, a source then regarded as uninteresting in chemistry. However, this minor constituent unexpectedly became an important factor leading to troponoid chemistry or what is now more broadly called nonbenzenoid or novel aromatic chemistry.

Furthermore, hinokitiol, which was obtained from unexpected factory waste after Japan's defeat in World War II, became an extremely precious compound for me. In addition, the thoughtful arrangements of the National Taiwan University authorities and the students' active cooperation were very important factors leading to the foundation of tropolone chemistry. Professor L. Crombie, in his paper[202] entitled "Shaping Chemistry Through the Study of Natural Products", referred to Kekulé's benzene structure, Dewar's proposed tropolone structure for stipitatic acid, and my independent proposal of a seven-membered structure for hinokitiol. After mentioning that the tropolone and hinokitiol structures led to the development of the novel area of nonbenzenoid aromatic chemistry, Crombie quoted seven names that he thought contributed most to organic chemistry. I read this paper and was quite flattered by his recognition of this special field in organic chemistry.

I discovered hinokitiol by "accident", and this led to the discovery of its very unique properties. If I had not paid attention to this very simple compound containing only 10 carbon atoms and two oxygen atoms, I might have achieved nothing of interest to the scientific community. Although I am not clear-headed, my curiosity made many future endeavors possible. Despite its simplicity, hinokitiol, or tropolones in general, becomes aromatic by 6π resonance contributions when the compound is planar and yet becomes polyolefinic when it adopts a nonplanar structure because of its ring strain. The uniqueness lies in the fact that this ring system becomes either aromatic or polyolefinic according to various structural factors and the reaction environment. I think that troponoids and the related compounds are excellent targets for physical, organic, and theoretical chemistry, as well as interesting compounds to be used as biochemical tools. I had accidentally encountered this very unique compound, and through this discovery, I have experienced deep satisfaction as a chemist. If I have been able to make an important contribution to organic chemistry, it is not because of my ability but, rather, because of the unique properties of troponoids and my curiosity for the unknown. In addition, I have received the continuous, warm support of the Kao Corporation and other companies for an unprecedentedly long time for my simple, basic research. I am quite a lucky chemist!

In retrospect, during my studies on hinokitiol and tropolone, I frequently encountered huge barriers caused by the exceptional properties of these compounds, which, unavoidably, led sometimes to incorrect structures and incorrect interpretations of reactions. Every time such a mistake was recognized by myself or a barrier was overcome by our later studies, the research progressed smoothly for a while. Then the next barrier would appear. This pattern was repeated several times. There is a saying, "Failure leads to success," which I have experienced frequently. About 15 years ago, a column in the Asahi Press (the most important newspaper in Japan) said the following: "The important thing in science is to fail fruitfully. Professor S. Tomonaga, Nobel laureate in physics, as he often said, was a master in this art." I must fully agree with this.

I have experienced numerous times the importance and significance of having a childlike curiosity on matters that appear to be trivial. When I succeeded in the synthesis of tropolone, hinokitiol, and other thujaplicins, an elder friend of mine kindly commented that because I had achieved the synthesis of the most important compounds in this field I should move on to another more important subject. However, I did not follow this advice; the unexpected reactivities and uniqueness of the hinokitiol reactions that I encountered in Taiwan led me to believe that completion of a synthesis is only the first step. The most important step is the clarification of hinokitiol's unusual properties. I learned that the development of a new field is a direct result of a passion for research and unending pursuit.

With respect to my first research topic, I realize that it was quite different from many others. It is often said that the research of a scientist is frequently governed by the topic given by one's first mentor. In my case, my undergraduate research topic at Tohoku Imperial University was the synthesis of the incorrect thyroxine structure. The reasons leading to polyterpenoid research at Taihoku Imperial University in Taiwan were also, as described earlier, very unusual. I believe that the first research topic plays an important role in determining the direction of one's future research, but more important is to find oneself a truly exciting topic that deserves to become one's life work.

Health is an important factor in a science like organic chemistry, where physical endurance is necessary. Considering my poor health when I was young, neither I nor my parents thought I would live to be

Tetsuo Nozoe as a gardener at his second house in Jogasaki, the Izu Peninsula, about 30 km southwest of Tokyo, August 1983. He goes there on weekends a few times a month to garden, keep house, or write, depending on the weather.

40 or 50. My recent health has been pretty good for an old man, and people tell me that I am "very young". However, this is not exactly true. (My family always says of me, not young but childish. I agree with it and am satisfied.) The origin of my health (at least my apparent health) is the result of other people's good will and my love for research and the unknown. In particular, at an old age, it is essential that one remains active physically and mentally. My residence in Tokyo is a "rabbit hutch", and there is very limited space for study or entertaining friends, particularly overseas friends. Fortunately I have a small second house on the Izu Peninsula, which is 3 to 4 hours away from Tokyo. I go there with materials for writing papers every weekend, if possible, and often alone. I look after the garden and plants but do little writing. This is an important factor in keeping me healthy and temporarily diverted from chemistry. Also, it is not healthy to dwell on failure. The important thing is to profit from it and to forget it. It is important to look forward with hope and to have the confidence to do some meaningful thing in science or for the welfare of human beings.

One of my weaknesses is a lack of knowledge in mathematics, physical chemistry, and writing, particularly in English. In my middle school days, I exchanged stamps with pen pals abroad, but we wrote in Esperanto. During the early stage of my research, I lacked international perspective, and most of my work was published in Japanese. Even after I realized the necessity of writing in English, I asked other people to translate. Today, this situation has not changed much, although in the past several years, I have managed to publish papers by writing them in English by myself first and then asking my former student, Professor H. Yamamoto, to check the manuscripts. It has been a major drawback for me. However, giving lectures abroad in English, on the basis of a prepared English manuscript is not a burden anymore, a result of both audience interest in our studies and my excitement in hearing about their recent research. Since my first overseas trip in 1953, I have come to truly realize the importance and the benefits of international exchange. Wherever I go, the kindest people are those involved in the same area of research or who are in some sense my competitors. Fortunately, science is a gracious field.

In retrospect, I begin to think that I could have done better. I expanded my studies in troponoids after realizing their unique properties. However, I was very slow in writing papers and review articles. In 1960, I, together with my research team, wrote a 700-page monograph on the chemistry of nonbenzenoid aromatic compounds.[94] Because the monograph was in Japanese, I was asked by the Elsevier Publishing Company to translate it into English. The announcement was already made in 1970. However, when I started to write, I felt that troponoid chemistry was totally incomplete and unbalanced, because the basic reactions of the fascinating troponoid system had been performed more than 40 years ago, in the pre-modern age, and not many people had carefully repeated the experiments since then. I thought I would write the monograph after updating the chemistry, and years have simply gone by, and no English monograph has been published.

At present, the total number of the published troponoid and azulenoid compounds already exceeds 10,000, and the number of papers, including theoretical calculations, exceeds 6000. I have finally realized that it is impossible to ask other busy, active professors to help me as coauthors in this time-consuming work. I feel strongly that it is my duty to pass on this information to the general chemistry community. To repay all the assistance and honors I have received, it is my duty to undertake this myself so far as my health lasts. Fortunately, the Kao Corporation assisted me in this program to write a comprehensive monograph, *Comprehensive Tropylium Chemistry*, and several of my friends (Breslow, Nakanishi, and the late Professor Hamao Umezawa) strongly

*Dr. Yoshio Maruta, president of the Kao Corporation, and Tetsuo Nozoe discuss the latter's plans for his book (*Comprehensive Tropylium Chemistry*) in Nozoe's office at the Kao Tokyo Research Laboratory in 1989.*

encouraged me to have it published. Moreover, Professors Murata, Oda, and Yamamoto kindly offered their help in consulting on the contents and checking the manuscript.

The number of troponoids and azulenoids has already reached 10,000. Their structures vary greatly, and their chemical properties differ from those of traditional chemical compounds. Unfortunately, even in advanced textbooks, only a few pages (in most cases, one or two pages) are spared for this subject. The lack of attention seems to be because few researchers are studying these compounds, and almost no important application of these compounds has been found. Therefore, as one who has been receiving support from many people for the study of this subject, I feel obligated to publish a comprehensive book on this subject.

As one of the oldest organic chemists, I would like to comment a little on recent Japanese organic chemistry. My opinion is necessarily superficial, because since my retirement more than 20 years ago, I have only been following the special field of nonbenzenoid aromatic chemistry.

Tetsuo Nozoe proposes a kampai *(toast) at the 4th International Conference on Organic Synthetic Chemistry in Tokyo in 1982. It is a custom in Japan that an elder usually proposes a kampai.*

Owing to the research carried out by other advanced countries, organic chemistry in Japan, in particular natural products chemistry, has gradually gained in appreciation internationally. It is encouraging to see that in recent years the influence of Japanese chemistry has increased greatly. Unlike myself, young Japanese chemists now routinely spend 1 to 2 years abroad in a major laboratory supported by various postdoctoral fellowships or as graduate students. It has now become common to discuss science with domestic scientists, as well as overseas counterparts. In my era, we were much more timid and neither asked questions nor voiced our opinions, especially with foreign speakers. Probably this timidity is due to a special old Japanese characteristic, but it has changed quite considerably since World War II.

Because of Japan's economic growth, the scientific equipment at various universities and, in particular, companies, has improved dramatically in recent years. Also, because of the Japanese *kôza* system, associate professors and instructors usually help the professor's research. Furthermore, a professor does not have to pay staff members and graduate students from his grants. Therefore, in certain cases, it is easier to cope with large projects in Japan than it is in other countries. Japanese students are also very diligent. There are several societies in the areas related to chemistry in Japan.

Since the end of the war, both the members of *kôza* and students in universities have increased considerably. Soon after the war, organic chemistry in our country was stimulated by a new generation of chemists exemplified by the following: Mukaiyama, Yoshida, Nozaki, and Noyori in organic synthesis using organometallics; Kametani and Ban in alkaloid synthesis; Matsui and Mori in synthesis of bioactive compounds; and Nakanishi in microscale structure determination, especially with respect to methodology. Many more have followed since. Many symposia in specialized areas are jointly organized by the various societies, and intense discussions are commonly held at these meetings. Also many of our young generation have gone abroad, joined the best groups, and returned with newly acquired knowledge. These factors have certainly contributed to the recent quantum leap in the level of organic chemistry and chemistry in general in Japan.

During my active years of work, the research was done mostly by one or two staff members (mostly without Ph.D.s) and several undergraduate students. Compared with this earlier time, the level of present research has improved, naturally. However, I also have a feeling that research may sometimes become somewhat shallow to secure research funds, accomplish neat work quickly, and produce Ph.D.s. Also, in the rush for positive results, I wonder whether some people are ignoring data or minor or "resinous" compounds that do not fit their expectations and, thus, lose leads for discoveries.

As I mentioned earlier, after the war, Professor Roger Adams visited Japan and encouraged and gave hope to young Japanese organic chemists. In 1962, it was beneficial for Japan that Professor Derek Barton toured Japan and excited young chemists by his brilliant lectures on stereochemistry and photochemistry. The 1964 3rd IUPAC Natural Products Symposium organized in Kyoto by Professors Munio Kotake and Kyosuke Tsuda was a big event, which very much contributed to the development of Japanese organic chemistry. Many major chemists attended from all over the world and inspired young Japanese chemists; the meeting also made overseas participants appreciate the level of Japanese organic chemistry.

The recent progress in Japanese organic chemistry is due mainly to influences from abroad. From now on we should strive to produce only truly original work of the highest standard and no longer be content with less than the best.

I am concerned that so many people have a tendency to handle the most fashionable topics and to attach little attention to more important basic problems. I also think that some people tend to take the word "originality" too casually. When I was young, we compensated for the lack of modern convenient equipment by improvisation. Either we or the technicians produced the required equipment. Nowadays,

Sir Derek Barton (Nobel Prize, 1969) is congratulated at the conclusion of his lecture at the Hitachi Family Center in Sendai, 1962. Nozoe was chairman of the event.

analytical and isolation equipment have advanced to such an extent that there is a tendency to rely on them too much and to neglect making one's own improvements. As for the situation of troponoid chemistry, because of a lack of more appropriate effort on our part and especially because of the unexpected properties of these compounds, unfortunately, or maybe fortunately, many important problems in this field remain to be clarified even today. Troponoid chemistry is still a developing area in organic chemistry. We still do not know how to control these complex reactions such that they proceed in one's desired direction.

Epilogue

My Autograph Books

The autographs throughout this volume are but a very small part of the collection I have acquired since I began this hobby in the 1950s. Between the pages of each of the 11 volumes I have accumulated are interesting comments, kind wishes, and fond memories of many exceptionally talented people. It was most difficult to select those few from so many wonderful moments. Because of space limitations, I had to edit the autographs by omitting structures and sketchs that accompanied many of these signatures.

I began my collection with my first trip abroad, thinking these signatures would be a special memento of my travels. When I saw the enthusiastic response of those I asked, and realized how much I enjoyed re-reading their comments, I decided to pursue my hobby more vigorously. I then began requesting autographs from friends and colleagues wherever and whenever we gathered—both at home and abroad—at meetings, lectures, labs, and social gatherings.

When I first began selecting which autographs to include in this book, I was hesitant to include many Japanese and Chinese contributions because most were written in those languages, and I was concerned that many readers would not truly appreciate their comments without lengthy explanation. Upon reflection, I realize that having had the good fortune not only to have met a number of world famous organic chemists of many *nationalities, but to have captured a few of their thoughts on paper, it would be remiss of me—and the reader's loss—not to share them.*

Therefore, I would like to take the opportunity of including here some very special reminiscences from my Japanese and Chinese friends and colleagues.

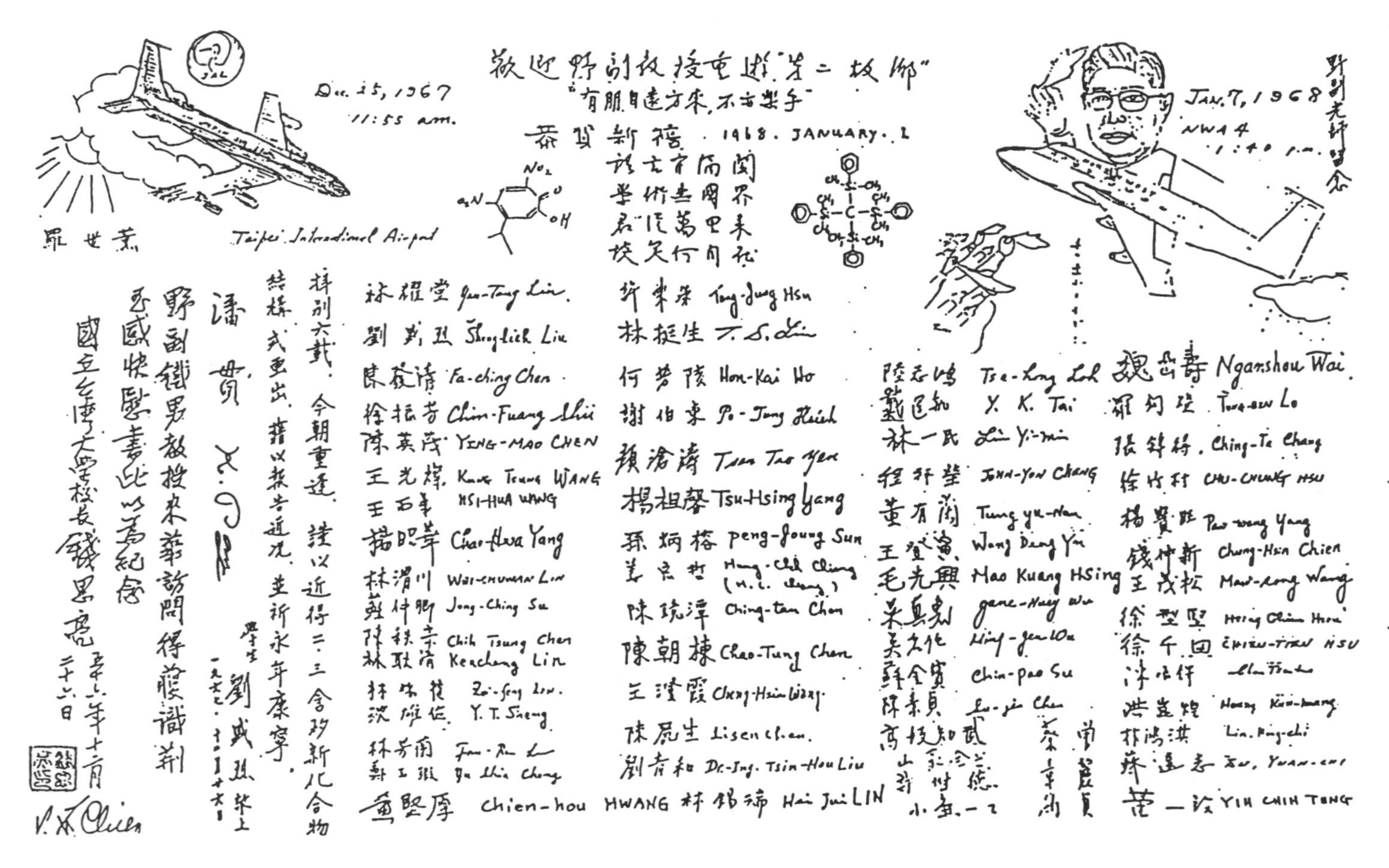
Dec. 25, 1967
11:55 am.
Taipei International Airport
1968. JANUARY. 2
JAN. 7, 1968
NWA 4
Ngamshou Wai
YING-MAO CHEN
Tsu-Hsing Yang
Peng-Joung Sun
Chao-Tung Chen
Lisen Chen
Mao Kuang Hsing
Chung-Hsin Chien
Chien-hou HWANG
Hai Jui LIN
YIN CHIN TONG
V. K. Chien

In 1967, Dr. Nozoe returned to Taiwan, 20 years after his repatriation to Japan. During that trip he gave a series of 5 lectures and was honored at banquets every day and evening. This reunion resulted in many drawings and over 100 signatures, of which 64 are shown here. The first line of the autograph centered at the head of the page says "We'd like to give Dr. Nozoe a hearty welcome on his return to his second home." The 3 lines on the left were from a welcoming speech by Dr. S. L. Chien, president of National Taiwan University, who discovered the biphenyl-type stereoisomers when working with Roger Adams. Sheng-Lieh Liu, an organosilicon chemist, contributed the next 4 lines of welcome on behalf of the graduates of Taihoku Imperial University, where Nozoe had taught. The drawings in the upper left and right corners were drawn by the 12-year-old son of Tung-Bin Lo and depict Nozoe's arrival in and imminent departure from Taipei (5 hours after the farewell party.) The tiny characters under the plane are Japanese for "until we meet again".

宋會端伯有花中十友 菊為逸友
桂為仙友 蓮為淨友 梅為清友
海棠名友 荼蘼韻友 瑞香殊友
芝蘭芳友 臘梅奇友 梔子禪友
私は貴君の何友でせうか
薬翁朝比奈泰彦 九十叟
昭和辛亥晩春

於 薬理研究会研究所
Dec 24, 1954 Y Asahina

In May 1971, Nozoe visited Yasuhiko Asahina at his office in Tokyo. The 90-year-old professor emeritus wrote an old Chinese poem in beautiful calligraphy. The gist of the poem is that there are 10 kinds of friends and each is likened to a type of fragrant flower. Asahina ended the poem with the reflection, "To you, I wonder what category of friend do I belong?" Asahina was the foremost natural products chemist in Japan at the time. He, along with Yuji Shibata and Riko Majima, had studied in the laboratory of Willstätter and Werner at the ETH, Zurich, around 1910. Shoji Shibata, Yuji's nephew, became Asahina's successor at the University of Tokyo. (The western version of Asahina's signature placed at the bottom of the poem was obtained much earlier, in 1954.)

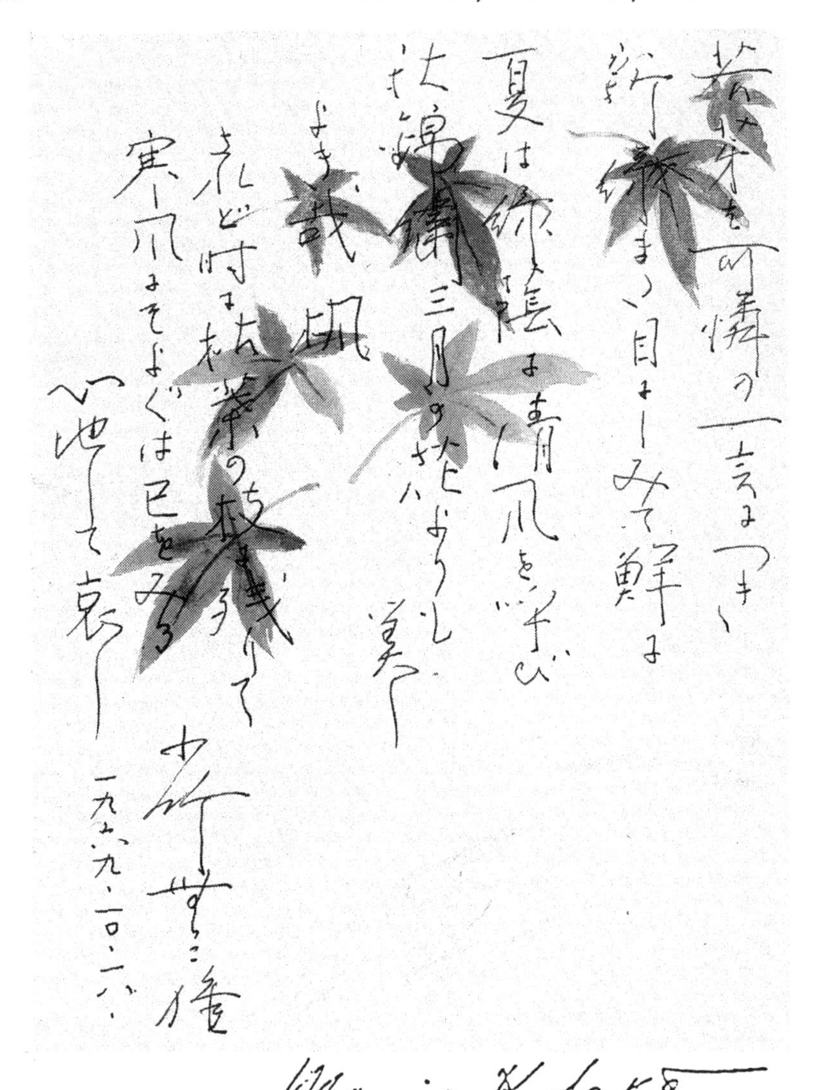

Munio Kotake

A Japanese poem written and illustrated by Munio Kotake was added to the collection on October 18, 1969. The maple leaves are delicately tinted in shades of summer green and the reds and golds of autumn. Professor Kotake was an intellectual in the traditional Japanese style, with many friends who shared his tastes. He was a first-class illustrator of Japanese plants in the traditional style and published a book of his beautiful drawings. (A comment from Nozoe: "His calligraphy is also first class, while the author's is third class.") Professor Kotake was a pioneer in natural products chemistry in Japan.

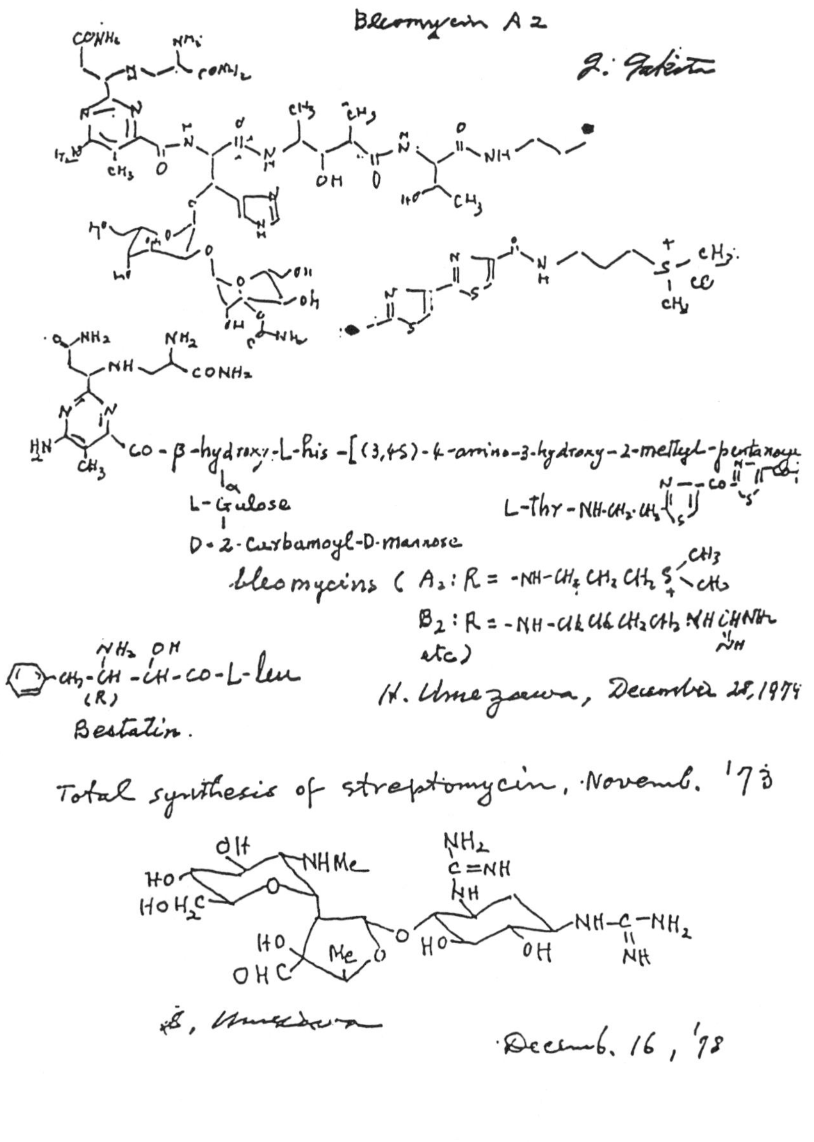

Top: a bleomycin structure from Dr. Tomihisa Takita. Below: structures and signatures of Hamao Umezawa, a medical chemist and close friend of the author, and his elder brother, Sumio, a synthetic chemist. Hamao discovered nearly 100 new antibiotics, determined their structures, and Sumio synthesized most of them, e.g., streptomycin, kanamycin, bleomycin.

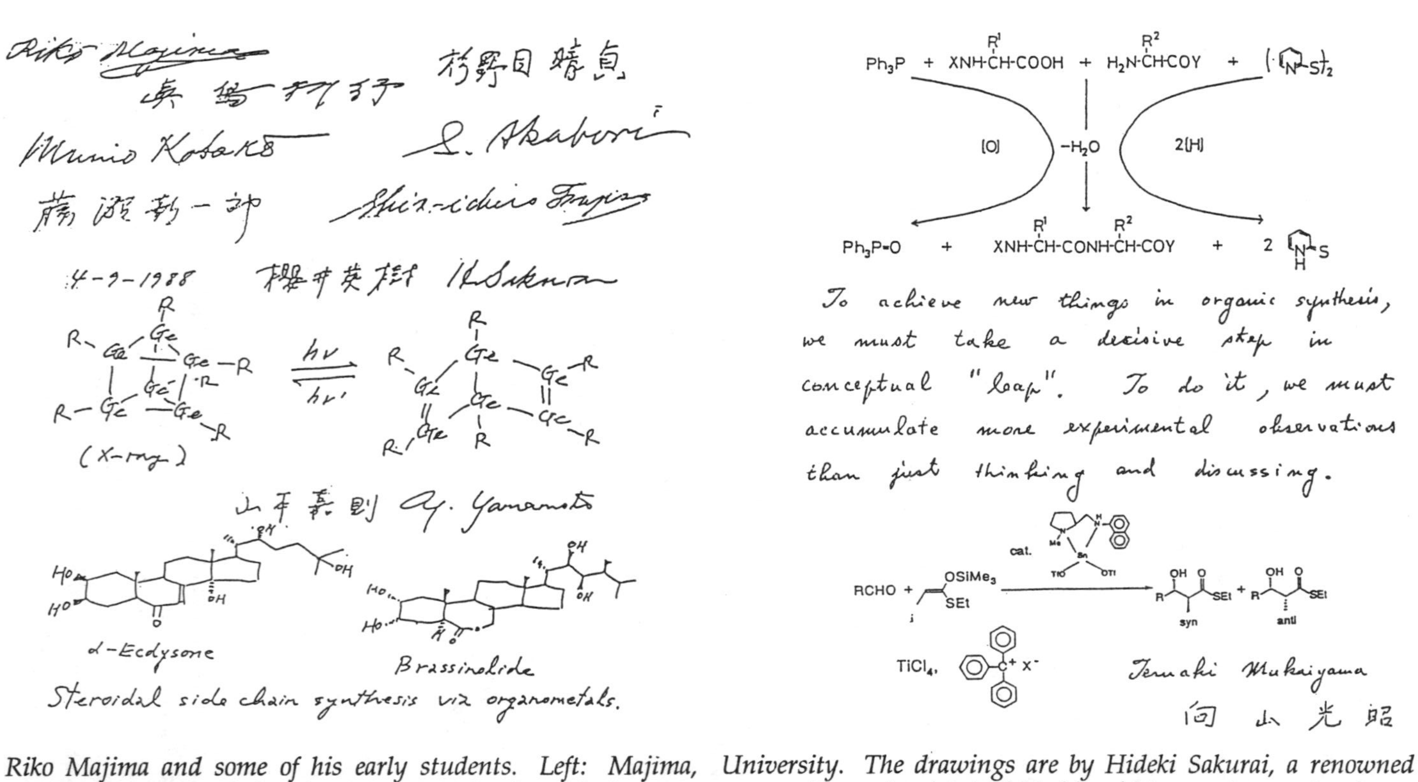

Riko Majima and some of his early students. Left: Majima, president of Osaka University (1943–1946), signed in Japanese and western. To his right, the Japanese signature of Harusada Suginome, president of Hokkaido University (1954–1966). Below, Munio Kotake, Shiro Akabori, president of Osaka University (1960–1966), and Shinichiro Fujise, Tohoku University. The drawings are by Hideki Sakurai, a renowned organosilicon chemist, and Yoshinori Yamamoto, a natural products and high-pressure chemist, both professors at Tohoku University. Right: Wise advice from Teruaki Mukaiyama, professor emeritus, Tokyo University, now with the Science University of Tokyo, and one of Japan's eminent organic chemists.

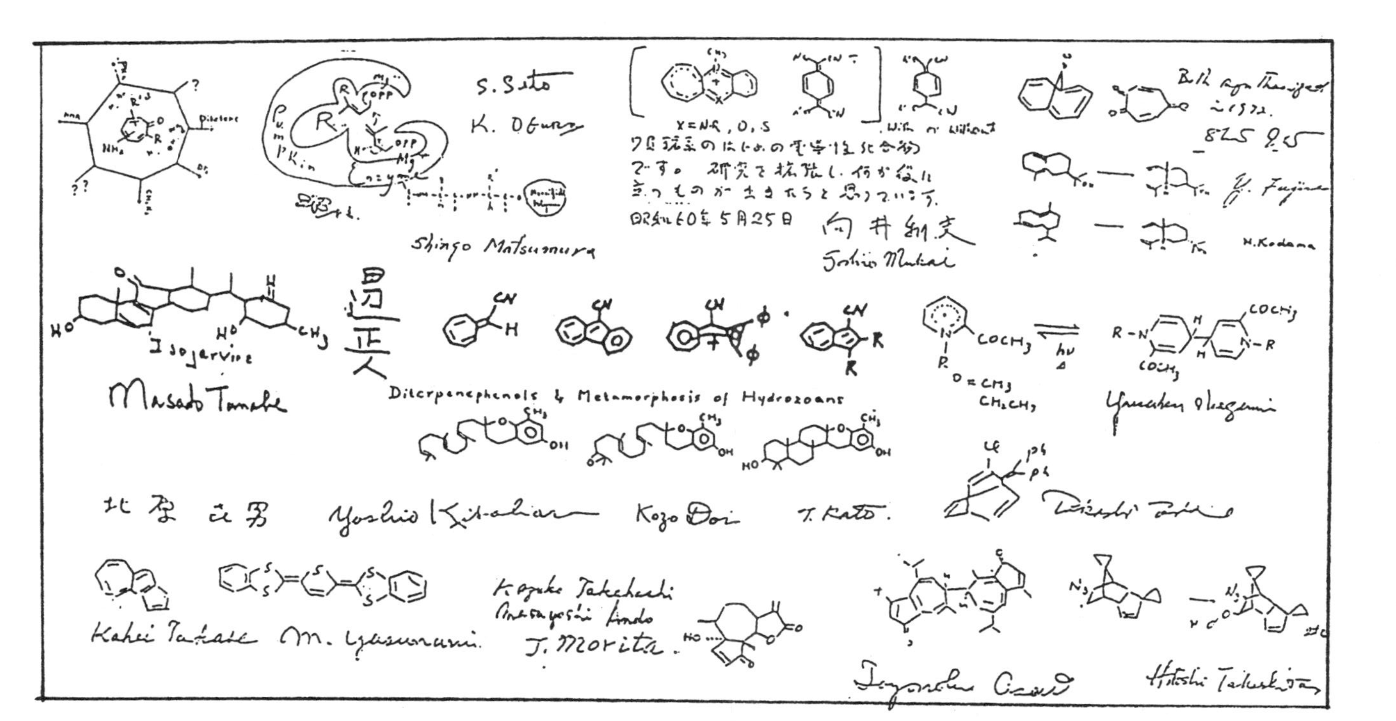

Some of the author's students and co-workers at Tohoku University. Beginning from the top left: Professors Shuichi Seto and Kyozo Ogura; Shingo Matsumura and Masato Tanabe (a Fulbright Scholar now at Stanford Research Institute); Yoshio Kitahara, Kozo Doi (Hirosaki University), Takahiro Kato (Science University of Tokyo); Toshio Mukai (Nihon University) and Shô Itô (Tokuyama Bunri University); Yutaka Fujise (Toyohashi Medical College), K. Kodama (Tokushima Bunri University) Kahei Takase, Masafumi Yasunami, Kazuko Takahashi, Masayoshi Ando (Niigata University), and Tadayoshi Morita; Yusaku Ikegami (director, Institute for Chemical Reaction Sciences); Toyonobu Asao; Takashi Toda (Utsunomiya University); and Hitoshi Takeshita (Kyusyu University).

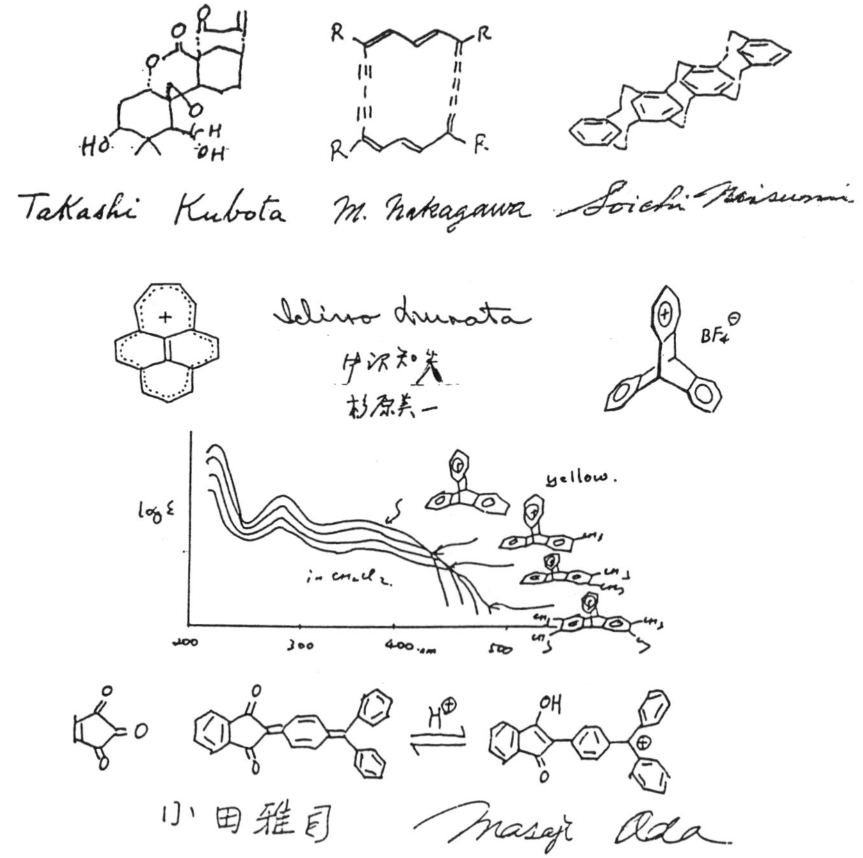

Some of Nozoe's colleagues and students from Osaka University: Professor Takashi Kubota, emeritus, Osaka City University, was Majima's last student at Osaka); Masazumi Nakagawa, a famous macroring annulene chemist; Shoichi Misumi, was a student of Munio Kotake; and Ichiro Murata; the Japanese signatures are those of Tomoo Nakazawa (Yamaguchi Medical College), Yoshikazu Sugihara, and Masaji Oda.

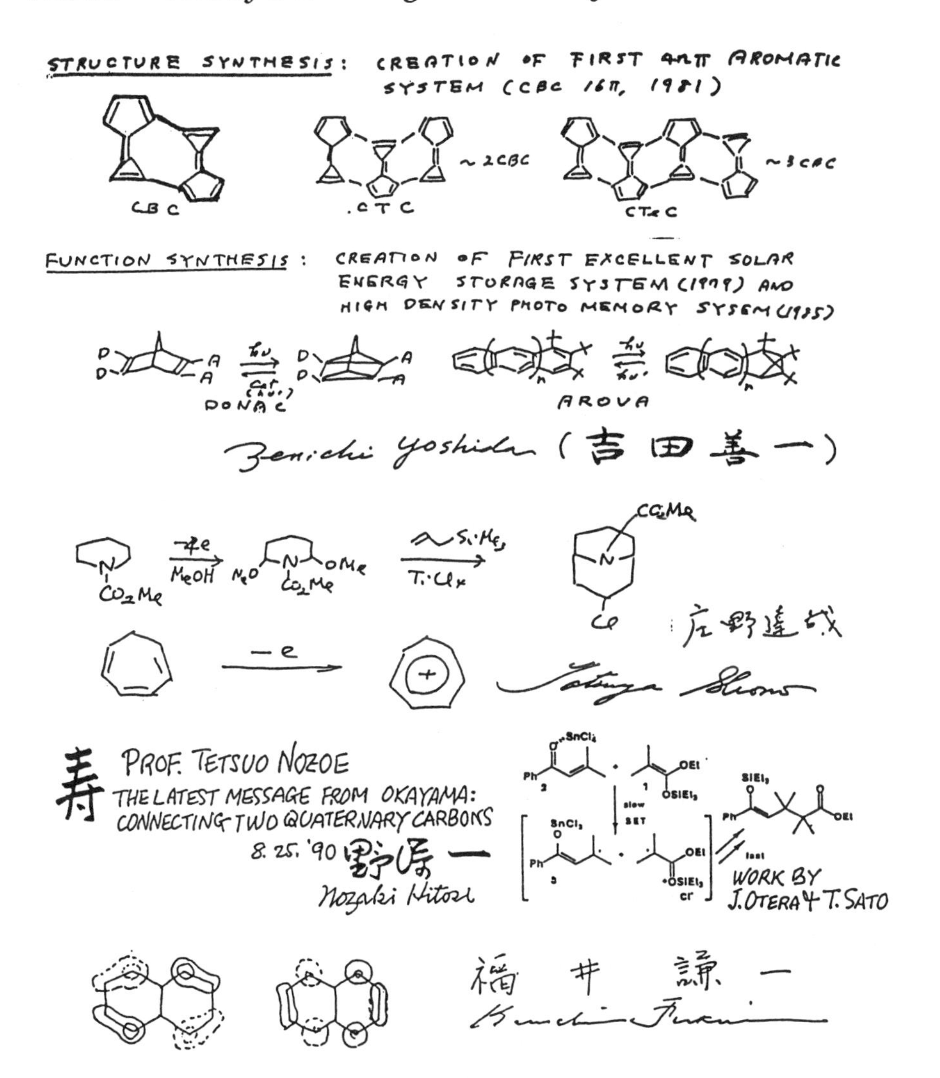

Contributions from Nozoe's colleagues at Kyoto University, Faculty of Engineering. Zenichi Yoshida (Kinki University), novel aromatic chemistry; Tatsuya Shono, electro-organic chemistry, who has been collaborating with Nozoe on the synthesis of tropenoid and azulenoid compounds by EOC. Hitoshi Nozaki, an eminent synthetic organic chemist (Okayama Science University), and Kenichi Fukui, director of the Institute for Basic Research, and only Japanese winner of the Nobel Prize for chemistry, 1981.

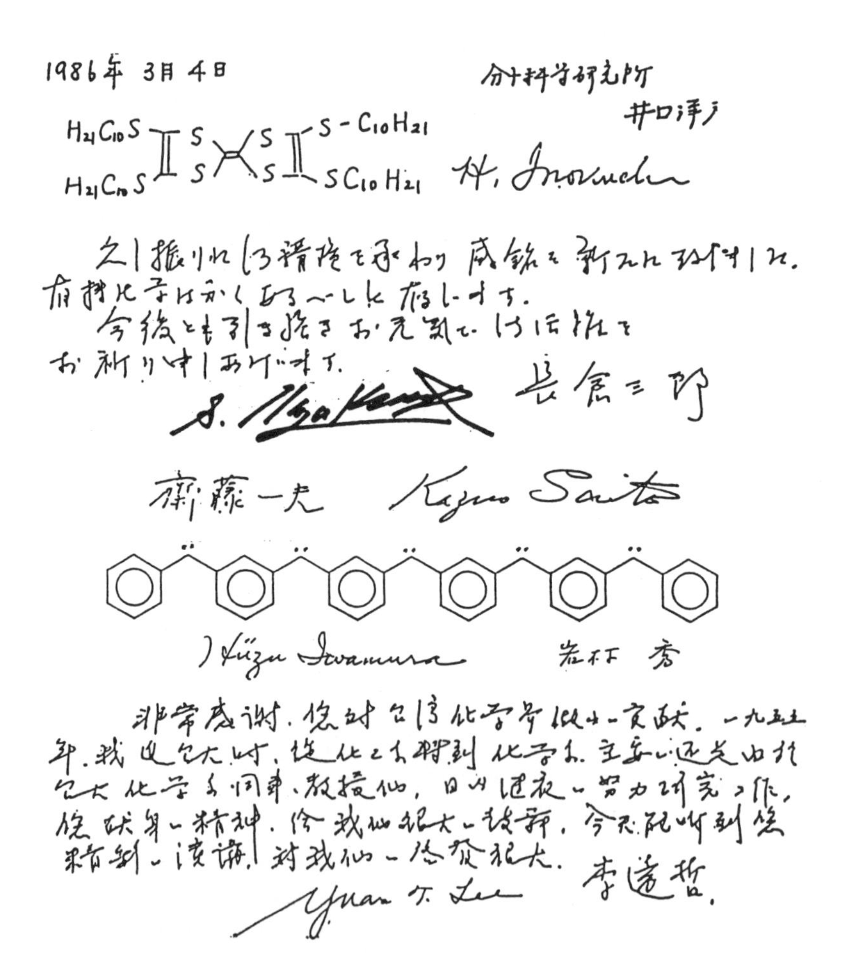

A memento of the author's visit to the Institute of Molecular Sciences in Okazaki, March 1986. Leading the list is the current director, Hiroshi Inokuchi, followed by Saburo Nagakura (director at the time of Nozoe's visit), Kazuo Saito (Tohoku University), Hiizu Iwamura (University of Tokyo), and another visiting scientist, Yuan T. Lee, who received the Nobel Prize a few months after this meeting. Lee, a student of one of Nozoe's students, commented in Chinese: "The reason I moved from the Faculty of Engineering to the Faculty of Science at Taiwan University is because I was impressed by the hard working professors who, as your former students, reflected the contributions of Professor Nozoe to Taiwan chemistry, for which I thank you very much."

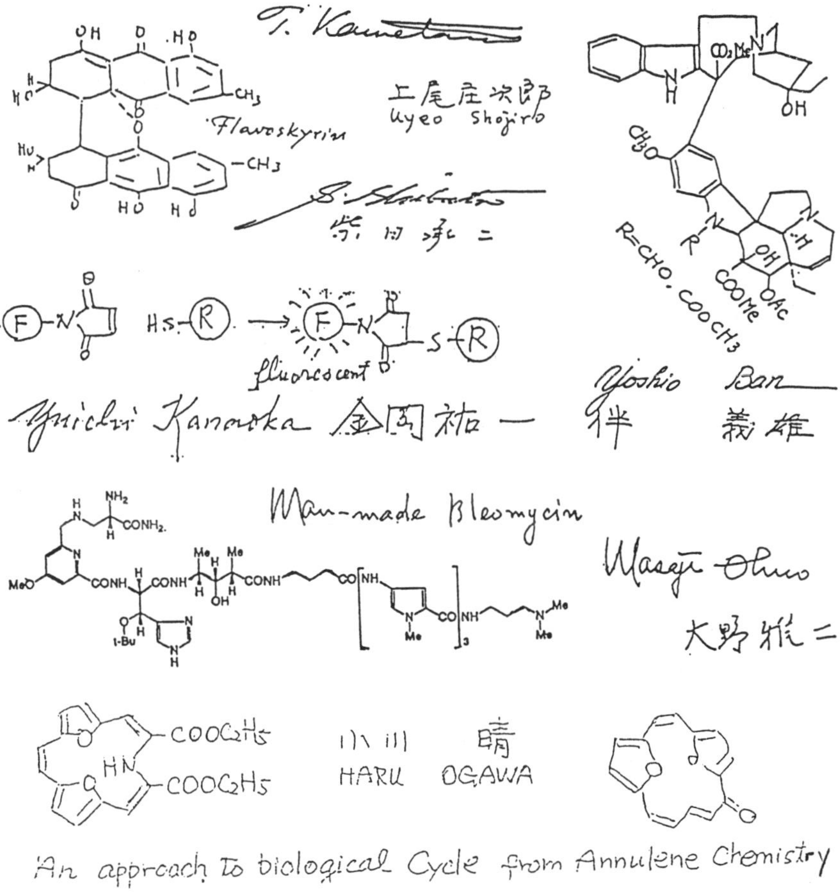

Organic chemists from several different faculties of Pharmaceutical Sciences, many of them natural products or synthetic chemists. Shoji Shibata, emeritus, Tokyo University, contributed a structure, as did Yoshio Ban (former president of Hokkaido University). Other signatures include Tetsuji Kametani (Tohoku and Hoshi Universities), Shojiro Ueo, (Kyoto University), Yuichi Kanaoka (Hokkaido University), Masaji Ohno (Tokyo University). Haru Ogawa (Kyushu University) is well known for his heteroannulene studies.

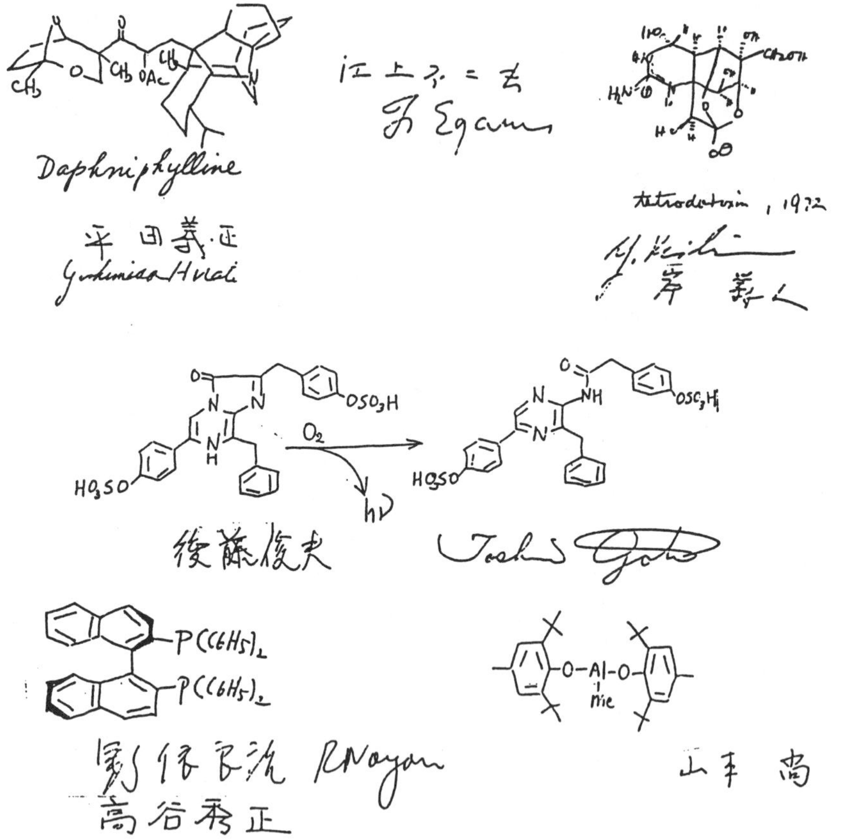

The staff of the Nagoya Egami School. Fujio Egami (the founder), biochemist; Yoshimasa Hirata, natural products chemist; Yoshito Kishi (Harvard), famous for his synthesis of complex natural products; Toshio Goto (deceased), natural pigments chemist. Koji Nakanishi (Columbia University) was also a member of this group. There are several well known researchers at Nagoya University: Ryoji Noyori, a synthetic organic chemist using asymmetric metallic catalysts; his co-worker Hedemasa Takaya (now at Kyoto University), and Hiroshi Yamamoto.

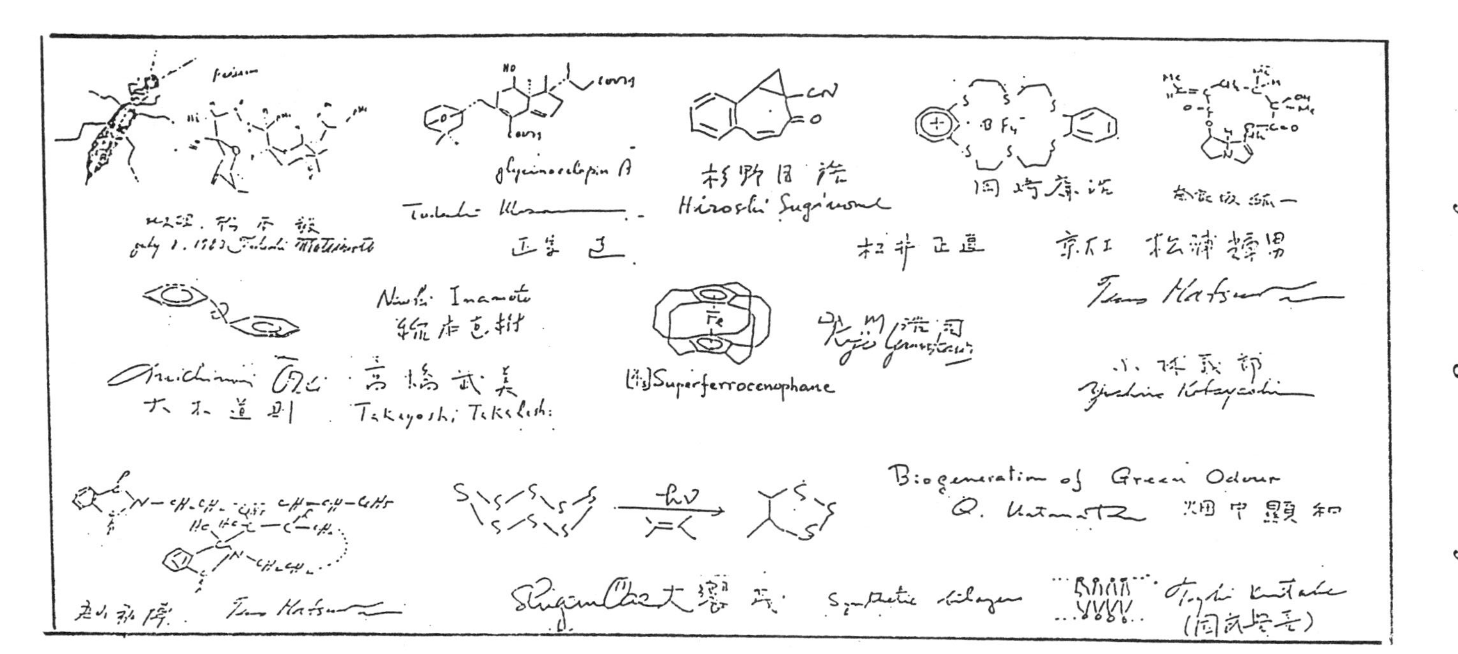

A few of my fellow organic chemists, some of whom are now retired: Osamu Simamura and Michinori Oki (Okayama Science University); Renji Okazaki, Naoki Inamoto, Takeyoshi Takahashi, and Kôichi Narasaka, (Tokyo University); Teruo Matsuura and Kazuhiro Maruyama (Kyoto University); Koji Yamakawa (Tokyo Science University); Yoshio Kobayashi, (Tokyo College of Pharmacy); Takeshi Matsumoto, Tadashi Masamune, and Hiroshi Suginome (Hokkaido University) drew the structure of pederin, a potent poison they had isolated from an insect, among other things; Shigeru Oae (Okayama Science University); Akikazu Hatanaka (Yamaguchi University); and Toyoki Kunitake (Kyusyu University).

"NIH-SHIFT

T, R [O] → OH, T, R

Nov. 25, 1966

A. YUKU WARE NI
TODOMARU NARE NI
AKI FUTATSU (BUSON)

Bernhard Witkop
Tom
Marlene Witkop

A.) Loosely translated, the lines read: "I am leaving, you are staying; this fall brings us a different experience." (from Shiki's haiku).

B.

SO KAN
SHI AMA-
SAI TTE
RAI KOE
I NASHI

HAKU-IN

C.

NOZOE SEN-SEI NO CHŌ-JU O
O-INORI MOSHI AGE SŌRŌ.
SHIMOTSUKI KITSU-JITSU SHŌWA GŌ-JU NEN
EDO NI OITE

B.) Witkop received the Second Order of the Sacred Treasure at the time Nozoe was president of the Chemical Society of Japan. At the banquet he signed Nozoe's book with this old Chinese poem about a man too moved to speak when reunited with his teacher. C.) He continues, "My sincere wishes to you for a long life. A felicitous day in November 1975, in Edo" (the old name for Tokyo).

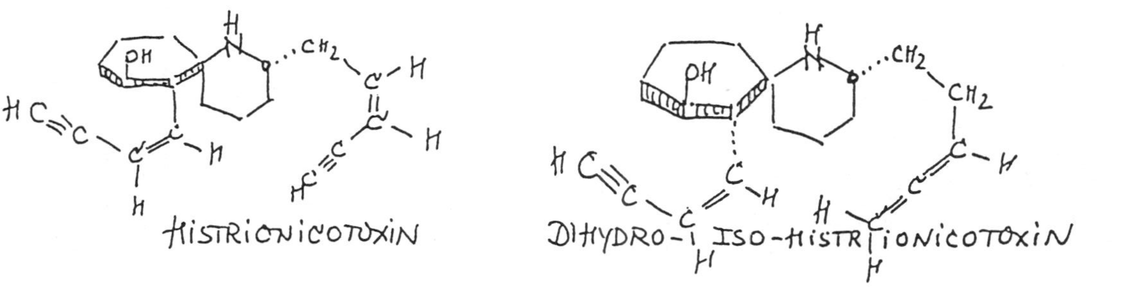

D.

TENTAKAKU UMA KOYURU NO KŌ
SOSHITE TOMO ARI EMPO YORI
KITARU TANOSHI KARAZUYA NO DE
DAI-AN-KICHI-JITSU TO
ZONJI SŌRŌ!

Bernhard Witkop

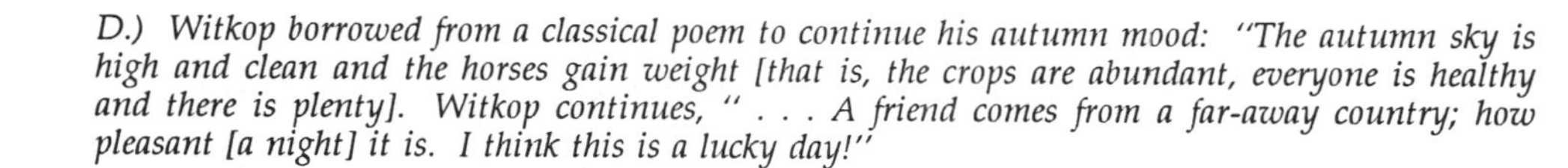

D.) Witkop borrowed from a classical poem to continue his autumn mood: "The autumn sky is high and clean and the horses gain weight [that is, the crops are abundant, everyone is healthy and there is plenty]. Witkop continues, " . . . A friend comes from a far-away country; how pleasant [a night] it is. I think this is a lucky day!"

A collection of some of the kind thoughts and greetings of friendship from Dr. Bernhard Witkop, acquired over 30 years. Witkop learned the Japanese language from Siro Senoh, who with Katsura Morita, was among the first of Japan's scientists to study at the National Institutes of Health. Witkop, a philosopher as well as a scientist, become very interested in both the classical language and culture of Japan. He frequently gives lectures in Japanese, but his spoken Japanese is "Shakespearean" Japanese and not always understood by the younger generation. His seal, which is pronounced as Witkop, is translated as "thin thread of rain in the old capital of Kyoto".

Roberts, Hammond, Winstein, Bartlett, Breslow
の各教授と一週間京都で過したことは長く忘れないでしょう
林 太郎

I shall never forget this one-week seminar in Kyoto spent with Professors Roberts, Hammond, Winstein, Bartlett, and Breslow. (T. Hayasi)

お蔭様でたのしいセミナー有難うございました
島村 修

Thank you for a wonderful seminar. (O. Simamura)

うまれて初めて1日9時間×4ぶっつづけで化学のディスカスするという又とない経験をさせてもらい生涯の記念になりました。YTBも国際的に発展させて戴きました。
湯川泰秀

We discussed chemistry 9 hours a day for 4 days! This was my first experience in such intense discussions, and it has become a precious memory for which I am thankful to you. I am also pleased to note that JTB (a nickname bestowed by Bartlett; it stands for "Japan Travel Bureau) has gradually become internationally appreciated. (Y. Yukawa)

アメリカですらこんなぜいたくなセミナールはあられない
ような大変な機会を作って頂き アメリカの超一流の physical
organic chemists のみならず日本の諸先生とも心から
うちとけた友人になれるチャンスを得まして非常に有難う
ございました

大饗 茂

Even in America it would be difficult to surpass the quality of this seminar. I am very grateful to you for giving us an opportunity to become well acquainted with the distinguished American physical organic chemists and with many prominent Japanese professors through our discussions. (S. Oae)

Impressions of the first Japan–U.S. Joint Seminar on Physical Organic Chemistry in Kyoto, April 1965, by 4 of the primary participants from the host country. Taro Hayasi (deceased), Ochanomizu University; Osamu Simamura, emeritus, Tokyo University and now with Sagami Research Institute; Yasuhide Yukawa, emeritus, Osaka University; and Shigeru Oae, emeritus, Tsukuba University and now with Okayama Science University. Their American counterparts were John D. Roberts, who organized the U.S. portion of the program, George S. Hammond (Caltech), Saul Winstein (UCLA), Paul D. Bartlett (Harvard), and Ronald Breslow (Columbia). Tetsuo Nozoe (Tohoku University) was the chairman of the seminar.

Excerpts from Nozoe's Address on the Occasion of his Beiju

The following is excerpted from Professor Tetsuo Nozoe's address to the participants at the symposium that took place on May 11, 1990, in Sendai, Japan, to honor Nozoe on the occasion of his Beiju *(88th birthday).*

Mr. Chairman, Ladies, and Gentlemen:

This is one of the most memorable days of my life. I am happy to be surrounded by many friends gathering to celebrate my 88th birthday and to discuss many aspects of organic chemistry all day long. I thank all of you for giving me such a wonderful day.

As one of the elder organic chemists in Japan, perhaps you will allow me to say a few words to the young people here today. I was very fortunate to have good teachers. Mr. Hasegawa, my middle school teacher, triggered my interest in chemistry. Professor Majima at Tohoku University and then Dr. Kafuku at the Central Research Institute of the Formosan government further strengthened it. These three people taught me the importance of simple curiosity, intellectual naiveté, and originality in thinking.

When I became an assistant professor at Taihoku Imperial University in 1929, I had to develop my own chemistry all alone because I was almost completely isolated from the rest of the world. Only three chemists with university training were in the whole of Taiwan when I moved there in 1926. This isolation helped me to recognize the importance of motivation to my research. It has had fundamental importance for my later research. If I had been in the center of Japan at that time, I could have benefited from easier access to the chemical literature and ready-made apparatus and also from frequent contact with other chemists. On the other hand, I would have been disturbed by all sorts of perturbations and would never have satisfied the curiosity, motivation, and naiveté I had. I would never have developed originality in my thinking. Therefore, I still recall the Taiwan period affectionately, when I spent many hours in the laboratory during and after the war. Incidentally, I visited Taiwan only a month ago and met many of my former students and technicians from Taihoku Imperial University and National Taiwan University.

If I have really been able to make any significant contribution to modern organic chemistry, it is not because of my ability but rather because of the unique properties of troponoid and azulenoid compounds and my curiosity for the unknown and even for minor resinous matters. In this sense, I always feel that I am quite a lucky chemist.

I will tell you also my weak point, which may give you a lesson. I have no research experience abroad. When I intended to go abroad, the war intensified and overseas travel became impossible. I did not try to write in English—please remember that we were under strong nationalism then—so most of my work was published in Japanese. I might be called an organic chemist genuinely "made in Japan." For many years, I lacked international perspective.

After the war, the situation in Japan changed dramatically. Since my first overseas trip in 1953, I have realized the importance and benefit of international exchange. I have made considerable effort to conquer my weakness in writing and speaking English. My conversation in English has improved to some extent, thanks to much more frequent contact with foreign chemists, but my writing remains unchanged. I still have to ask someone to improve my bad English. So I urge you, young Japanese students, to make the utmost effort to improve your English so that you may be able to exchange your opinions freely.

I express my gratitude to the late Professor Riko Majima and many senior organic chemists in Japan and abroad for their everlasting encouragement and kind help, to many more friends for their warm-heartedness, to my numerous co-workers for their devoted research for the past 60 years, and to my wife for her never-failing understanding.

A family gathering at the Nozoe home in Tokyo, October 1968. Front row from left: Yoko (Mrs. Ishikura), Tetsuo Nozoe, Kumiko (Yoko's daughter), Sumito (Yoko's son), and Satoru Masamune. Back row from left: K. Higashihara (Yuriko's husband), Yuriko, Kyoko, Takako (Mrs. Masamune), Tohoru (Takako's son), Hiroko (Takako's daughter), and Shigeo.

I express my sincere thanks to Takasago Perfumery Company,* Sankyo Company, and Kao Corporation for 40–60 years of unprecedented assistance. Finally, I thank you all again for this wonderful symposium and this memorable day. As I am still doing some small research and attending chemical meetings as frequently as possible, I sincerely hope to see you and talk to you many times. Thank you very much.

**Now the Takasago International Corporation.*

Acknowledgments

As I look back, I'm utterly amazed by the warm considerations extended by so many friends and colleagues. So many people have helped me that I can not mention all of their names here. The fact that my research has developed to this extent depended on many factors, including not only the understanding of the government, university, and colleagues (in Japan and Taiwan) but also of personal friends who recognized the importance of this research.

Furthermore, the financial assistance from Takasago International Corporation, Sankyo (Pharmaceutical) Company, and especially Kao Corporation is totally unprecedented. They have assisted me during the past 30–50 years independently of the company's profit and strictly for the sake of promoting basic research. For the past few years, the Japan Academy has also given me financial support to promote my cooperative work.

Most important is, without saying, the devoted work of numerous co-workers, technicians, and secretaries. Professors Koji Nakanishi and Hiroshi Yamamoto and Dr. Yuji Hazeyama (Myu Research, Tokyo) have greatly helped me with English for this chapter, and I thank my secretary, Mrs. Masako Doi, for typing and many other necessary arrangements. Moreover, Dr. Jeffrey I. Seeman, editor of this series, and the staff of the ACS Books Department have given me continuous encouragement and help in writing this book, and many of my former students have cooperated with me to identify names in old pictures. All of this support has been a solid encouragement for my continued work.

In ending, I want to express my sincere thanks to numerous colleagues, the government, universities, and companies who have helped my career over the past 60 years.

It is my honor to dedicate these memoirs to the two individuals who exerted an important influence on my life, Professor Riko Majima and Dr. Kinzo Kafuku.

References

1. Kendal, E. C.; Osterberg, A. E. *J. Biol. Chem.* **1920,** *40,* 265.
2. Nozoe, T. (a) *Proc. Imp. Acad. (Tokyo)* **1926,** *2,* 541; (b) *Nippon Kagaku Kaishi* **1927,** *48,* 147.
3. Harrington, C. R. *Biochem. J.* **1926,** *20,* 293, and following papers.
4. Kafuku, K.; Nozoe, T.; Hata, C. *Bull. Chem. Soc. Jpn.* **1931,** *6,* 40.
5. Kafuku, K.; Nozoe, T. *Bull. Chem. Soc. Jpn.* **1931,** *6,* 111.
6. Nozoe, T. *Nippon Kagaku Kaishi* **1934,** *55,* 1106, and following papers.
7. Nozoe, T. *Nippon Kagaku Kaishi* **1934,** *55,* 746.
8. Nozoe, T.; Kinugasa, T. *Nippon Kagaku Kaishi* **1935,** *56,* 883.
9. Haworth, R. D. *Annu. Rep. Prog. Chem. 1937* **1938,** *34,* 327–342.
10. Ruzicka, L.; Hofman, K. *Helv. Chim. Acta* **1937,** *20,* 1155, and papers cited therein.
11. Nozoe, T.; Katsura, S. *Nippon Kagaku Kaishi* **1937,** *58,* 1414.
12. Nozoe, T.; Kinugasa, T. *Nippon Kagaku Kaishi* **1937,** *58,* 589 and 590.
13. Nozoe, T.; Kinugasa, T. Presented at the 60th Annual Meeting of the Chemical Society of Japan, Tokyo, 1938; Abstract, *Nippon Kagaku Kaishi* **1938,** *59,* 772, 774.
14. Nozoe, T.; Katsura, S. Presented at the 60th Annual Meeting of the Chemical Society of Japan, Tokyo, 1938; Abstract, *Nippon Kagaku Kaishi* **1938,** *59,* 773.
15. Ruzicka, L.; Schellenberg, H. *Helv. Chim. Acta* **1937,** *20,* 1553.
16. Van der Haar, A. W. *Anleitung zum Nachweis, zur Trennung und Bestimmung der reinen und aus Glukosiden usw. erhaltenen Monosacharide und Aldehydsäuren;* Gebruder Borntraeger: Berlin, 1920; p 345.
17. Nozoe, T. *Nippon Kagaku Kaishi* **1935,** *56,* 852; **1936,** *57,* 798 and 813.
18. See, for example, Dareé, C.; Petrov, V. A. *J. Chem. Soc.* **1936,** 1562.
19. Windaus, A.; Tschesche, R. *Z. Physiol. Chem.* **1930,** *190,* 51.
20. Wieland, H.; Pasedach, H.; Ballauf, A. *Liebigs Ann. Chem.* **1937,** *529,* 68.

21. Kuwata, T.; Ishii, J. *Kogyo Kagaku Zasshi,* supplement binding, **1936,** *39,* B317, and references cited therein.

22. Abraham, E. E. U.; Hilditsch, T. P. *Chem. Ind. (London)* **1933,** *54,* T398, and references cited therein.

23. Nozoe, T.; Katsura, S.; Liu, S. L. Presented at the 61st Annual Meeting of the Chemical Society of Japan, Tokyo, 1939; Abstract, *Nippon Kagaku Kaishi* **1939,** *60,* 486; cf. Nozoe, T. Jpn. Patent 147,677, 1941, and other patents.

24. Nozoe, T.; Nakagawa, K.; Liu, S. L. Presented at the 61st Annual Meeting of the Chemical Society of Japan, Tokyo, 1939; Abstract, *Nippon Kagaku Kaishi* **1939,** *60,* 486.

25. The results of our study of wool wax acids are described later by S.-L. Liu, "Studies in the Fatty Acids from Lanolin", in *Essays and Papers in Memory of Late President Fu Ssu-nien,* National Taiwan University, 1952; pp 1–15.

26. Nozoe, T.; Katsura, S.; Yeh, P.-Y. Presented at the 61st Annual Meeting of the Chemical Society of Japan, Tokyo, 1939; Abstract, *Nippon Kagaku Kaishi* **1939,** *60,* 487; cf. Nozoe, T.; Jpn. Patents 153,628 and 153,629, 1942; *Chem. Abstr.* **1941,** *43,* 3219.

27. Nozoe, T.; Katsura, S.; Matsumura, H. Presented at the 59th General Meeting of the Pharmaceutical Society of Japan, Tokyo, 1939; Abstract 123; cf. Nozoe, T.; Jpn. Patent 147,100, 1941.

28. Marker, R. E.; Wittle, E. L.; Mixon, L. W. *J. Am. Chem. Soc.* **1937** *59,* 1368, and literature cited therein.

29. Nozoe, T.; Matsumura, H. Presented at the 63rd Annual Meeting of the Chemical Society of Japan, Tokyo, 1941; Abstract, pp 2, 3.

30. Nozoe, T.; Katsura, S.; Matsumura, H. Presented at the General Meeting of the Pharmaceutical Society of Japan, Tokyo, 1939; Abstracts 124, 125.

31. Tsuchihashi, R.; Tasaki, S. *Pept. Gov. Res. Inst., Dept. Ind. Formosa* **1924,** *5,* 119.

32. Hirao, N. *Nippon Kagaku Kaishi* **1926,** *47,* 666, 743.

33. Kawamura, J. *Bull. Imp. For. Exp. Stn. Jpn.* **1930,** *30,* 59.

34. Nozoe, T. *Bull. Chem. Soc. Jpn.* **1936,** *11,* 295.

35. Nozoe, T.; Katsura, S. Presented at the 60th Annual Meeting of the Pharmaceutical Society of Japan, Tokyo, 1940; Abstract 54.

36. Katsura, S.; Nozoe, T. and co-workers *Stud. Med. Trop., Taihoku Imp. Univ.* **1941,** *40,* 1557–1628; *Suppl.* **1944,** *1,* 1–352, and papers cited therein.

37. Pauling, L. *The Nature of the Chemical Bond;* Cornell University: Ithaca, NY, 1939.

38. Nozoe, T.; Katsura, S. *Yakugaku Zasshi* **1944,** *64,* 181.

39. (a) Nozoe, T. *Yakugaku* **1949,** *3,* 174–198; (b) English translation of ref. 39a with some additional material: Nozoe, T. *Sci. Rep. Tohoku Univ., Ser. 1* **1950,** *34,* 199–236.

40. Nozoe's isopropylcycloheptatrienolone structure 33 for hinokitiol was referred to by his co-worker: Katsura, S. *Saishin Igaku* **1947,** *2,* 595; *Medical Times* **1948,** *3,* 29 (presented at Formosan Med. Assoc., Taipei, 1947).

41. Nozoe, T. *Sci. Rep. Tohoku Univ., Ser. 1* **1952,** *36,* 40–62 and 82–98; Sebe, E. ibid. 99–105, 106–113, and 114–129; Nozoe, T.; Sebe, E.; Mayama, S.; Iwamoto, S. ibid. 184–202; Nozoe, T.; Sebe, E.; Kitahara, Y.; Fujii, H. ibid. 290–298; Nozoe, T.; Sebe, E.; Cheng, L. C.; Mayama, S.; Hsü, T. J. ibid. 299–306; Sebe, E.; Nozoe, T.; Yeh, P. Y.; Iwamoto, S. ibid. 307–322.

42. Erdtman, H.; Gripenberg, J. *Nature (London)* **1948,** *161,* 719; *Acta Chem. Scand.* **1948,** 2, 625.

43. Iinuma, H. *Nippon Kagaku Kaishi* **1943,** *64,* 742; *Chem. Abstr.* **1947,** *41,* 4731h.

44. Iinuma, H. *Nippon Kagaku Kaishi* **1943,** *64,* 901; *Chem. Abstr.* **1947,** *41,* 4731g.

45. Dewar, M. J. S. *Nature (London)* **1945,** *155,* 50, and following papers.

46. Dewar, M. J. S. *Nature (London)* **1945,** *155,* 141, 479.

47. Arnstein, H. R. V.; Tarbell, D. S.; Scott, G. P.; Huang, H. T. *J. Am. Chem. Soc.* **1949,** *71,* 2448; King, M. W.; De Fries, J. L.; Pepinski, R. *Acta Cryst.* **1952,** *5,* 437.

48. Aulin-Erdtman, G. *Acta Chem. Scand.* **1950,** *4,* 1061, and following papers.

49. Barltrop, J. A.; Nicholson, J. S. *J. Chem. Soc.* **1948,** 116.

50. Haworth, R. D.; Moore, B. P.; Pauson, P. L. *J. Chem. Soc.* **1948,** 1045.

51. Kitazato, Z. *Nippon Kagaku Kaishi* **1939,** *60,* 1055.

52. Ruzicka, R.; Grob, A.; van der Sluys-Veer, F. C. *Helv. Chim. Acta* **1939,** *22,* 788, and preceding papers.

53. Picard, C. W.; Sharples, K. S.; Spring, F. S. *J. Chem. Soc.* **1939,** 1045.

54. Spring, F. S. *Annu. Rep. Prog. Chem. 1940* **1941,** 191–203.

55. Noller, C. R. *Annu. Rev. Biochem.* **1945,** *14,* 382.

56. Tsuji, J.; Ohno, K. *J. Am. Chem. Soc.* **1967,** *90,* 94.

57. Barton, D. H. R. In *The Chemistry of Carbon Compounds;* Rodd, G. M., Ed.; Elsevier: Amsterdam, 1953; Vol 2B, pp 716–764.

58. Weitkamp, A. W. *J. Am. Chem. Soc.* **1945,** *67,* 447.

59. Truter, E. V. *Wool Wax—Chemistry and Technology;* Cleaver-Hume: London, 1956; pp 33–60, 236–318.

60. Warth, A. H. *The Chemistry and Technology of Waxes;* Rheinhold: New York, 1960; pp 122–341.

61. Doree, C.; McGee, J. F.; Kurzer, F. *J. Chem. Soc.* **1948,** 988.

62. Voser, W.; Montavon, M.; Gunthard, H. H.; Jeger, O.; Ruzicka, L. *Helv. Chim. Acta* **1950,** *33,* 1893.

63. Barton, D. H. R.; Fawcett, J. S.; Thomas, B. R. *J. Chem. Soc.* **1951,** 3147.

64. Nozoe, T. *Fortschr. Chem. Org. Naturst.* **1956,** *13,* 232–301.

65. Nozoe, T.; Kitahara, Y.; Yamane, K.; Yamaki, K. *Proc. Jpn. Acad.* **1950,** *26*(8), 14.

66. Nozoe, T.; Kitahara, Y.; Kunioka, E.; Doi, K. *Proc. Jpn. Acad.* **1950,** *26*(9), 38.

67. Nozoe, T.; Seto, S.; Kitahara, Y.; Kunori, M.; Nakayama, Y. *Proc. Jpn. Acad.* **1950,** *26*(7), 38.

68. Nozoe, T.; Seto, S.; Kikuchi, K.; Mukai, T.; Matsumoto, S.; Murase, M. *Proc. Jpn. Acad.* **1950,** *26*(7), 43, and following papers.

69. Nozoe, T.; Kitahara, Y.; Itô, S. *Proc. Jpn. Acad.* **1950,** *26*(7), 47.

70. Nozoe, T.; Seto, S.; Kikuchi, K.; Takeda, H. *Proc. Jpn. Acad.* **1951,** *27,* 146.

71. (a) Cook, J. W.; Sommerville, A. R. *Nature (London)* **1949,** *183,* 410; (b) Nakazaki, M. *Nippon Kagaku Kaishi* **1950,** *72,* 739; Sakan, T.; Nakazaki, M. *J. Inst. Polytech. Osaka City Univ.* **1950,** *1,* 23.

72. (a) Nozoe, T. *Proc. Jpn. Acad.* **1950,** *26*(9), 30; (b) Nozoe, T. Presented at the 2nd Annual Meeting of the Chemical Society of Japan, Kyoto, 1949; Abstract, p 27; Nozoe, T.; Kitahara, Y. C. O.; Abstract, p 27.

73. Kurita, Y.; Nozoe, T.; Kubo, M. *Nippon Kagaku Zasshi* **1950,** *71,* 543; *Bull. Chem. Soc. Jpn.* **1951,** *24,* 10, 13, and 99.

74. Doering, W. von E.; Knox, L. H. *J. Am. Chem. Soc.* **1950,** *72,* 2305, and following papers.

75. Kubo, M.; Nozoe, T.; Kurita, Y. *Nature (London)* **1951,** *167,* 688.

76. Cook, J. W.; Gibb, A. R.; Raphael, R. A.; Somerville, A. R. *Chem. Ind. (London)* **1950,** 427, and following papers.

77. Haworth, R. D.; Hobson, J. D. *Chem. Ind. (London)* **1950,** 441, and following papers.

78. *Tropolones and Allied Compounds;* The Chemical Society Symposium, London, *Chem. Ind. (London)* **1951,** *12,* 28.

79. Nozoe, T. *Nature (London)* **1951,** *167,* 1055–1060.

80. Cook, J. W.; Loudon, J. D. *Q. Rev.* **1951,** *5,* 99–130.

81. Huber, G. *Angew. Chem.* **1951,** 501–524.

82. Johnson, A. W. *Sci. Prog.* **1951,** *39,* 495–502.

83. Chopin, J. *Bull. Soc. Chim. Fr.* **1951,** 52D–57D.

84. Birch, A. J. *Annu. Rep. Prog. Chem. 1951* **1952,** *48,* 185–190.

85. Pauson, P. L. *Chem. Rev.* **1955,** *55,* 9–135.

86. For example, see Stevens, H. C.; Reich, D. A.; Fountain, K. R.; Gaugh, E. J. *J. Am. Chem. Soc.* **1965,** *87,* 5257.

87. Nozoe, T. *Croat. Chem. Acta* **1957,** *29,* 207–227.

88. Nozoe, T. *Experientia Suppl. 7,* **1957,** 306–327.

89. Nozoe, T. "Tropones and Tropolones", In *Non-Benzenoid Aromatic Compounds;* Ginsburg, D., Ed.; Interscience: New York, 1959; Chapter 7, pp 339–463.

90. Nozoe, T. *Pure Appl. Chem.* **1971,** *28,* 239–280.

91. Pietra, F. *Chem. Rev.* **1973,** 293–364.

92. Nozoe, T.; Murata, I. *MTP Int. Rev. Sci. Ser. 1* **1973,** *3,* 201–235; *MTP Int. Rev. Sci. Ser.* 2 **1976,** 197–228.

93. (a) Bertelli, D. N.; (b) Itô, S.; Fujise, Y.; (c) Mukai, T.; (d) Sugimura, Y.; Kawamoto, T.; Kishida, Y. In *Topics in Nonbenzenoid Aromatic Chemistry*; Nozoe, T.; Breslow, R.; Hafner, K.; Itô, S.; Murata, I., Eds.; Hirokawa: Tokyo, 1973; Vol. 1, (a) pp 29–46; 1976, Vol. 2, (b) pp 91–137, (c) pp 183–217, (d) pp 219–242.

94. Nozoe, T.; Takase, K; Matsumura, H.; Asao, T.; Kikuchi, K.; Itô, S. *Nonbenzenoid Aromatic Compounds,* Vol. 13, In *Dai Yukikagaku (Comprehensive Organic Chemistry)*; Kotake, M., Ed.; Asakura Shoten: Tokyo, 1960; Chapters 1–30, pp 1–691.

95. Lloyd, D. *Nonbenzenoid Conjugated Carboxylic Compounds;* Elsevier: Amsterdam, 1984; pp 1–431.

96. Asao, T.; Oda, M. *Methoden Org. Chem. (Houben-Weyl)* **1985,** *V/2c,* 49–86.

97. Asao, T.; Oda, M. *Methoden Org. Chem. (Houben-Weyl)* **1985,** *V/2c,* 710–789.

98. Becker, G.; Kolshorn, H. *Methoden Org. Chem. (Houben-Weyl)* **1985,** *V/2c,* 418–461.

99. Zeller, K.-P. *Methoden Org. Chem. (Houben-Weyl)* **1985,** *V/2c,* 127–417.

100. Nozoe, T.; Ikemi, T.; Ozeki, T. *Proc. Jpn. Acad.* **1955,** *31,* 455.

101. Nozoe, T.; Oyama, M.; Kikuchi, K. *Bull. Chem. Soc. Jpn.* **1963,** *36,* 168.

102. Nozoe, T.; Sato, M.; Itô, S.; Matsui, K.; Ozeki, T. *Proc. Jpn. Acad.* **1954,** *30,* 599, and following papers.

103. Masamune, S.; Kemp-Jones, A. V.; Green, J.; Rabenstein, D. L.; Yasunami, M.; Takase, K.; Nozoe, T. *J. Chem. Soc., Chem. Commun.* **1973,** 283.

104. Doering, W. von E.; Detert, F. L. *J. Am. Chem. Soc.* **1951,** *73,* 876.

105. Dauben, H. J.; Ringold, H. J. *J. Am. Chem. Soc.* **1951,** *73,* 876.

106. Nozoe, T.; Kitahara, Y.; Ando, T.; Masamume, S. *Proc. Jpn. Acad.* **1951,** *27,* 415.

107. Nozoe, T.; Mukai, T.; Takase, K.; Nagase, T. *Proc. Jpn. Acad.* **1952,** *28,* 477.

108. Kurita, Y.; Seto, S.; Nozoe, T.; Kubo, M. *Bull. Chem. Soc. Jpn.* **1953,** *26,* 272.

109. Nozoe, T.; Mukai, T.; Nagase, T.; Toyooka, T. *Bull. Chem. Soc. Jpn.* **1960,** *33,* 1247.

110. Nozoe, T.; Seto, S.; Takeda, H.; Morosawa, S.; Matsumoto, K. *Proc. Jpn. Acad.* **1951,** *27,* 556, and following papers.

111. Nozoe, T.; Seto, S.; Sato, T. *Proc. Jpn. Acad.* **1954,** *30,* 473. See also Haworth, R. D.; Tinker, R. B. *J. Chem. Soc.* **1955,** 911; Seto, S.; Sato, T. *Proc. Jpn. Acad.* **1954,** *30,* 473. The term ciné was first used in Bunett, J. F.; Zabier, R. *Chem. Rev.* **1951,** *49,* 273.

112. Kitahara, R.; Nozoe, T. Presented at the Symposium on Reaction Mechanism, The Chemical Society of Japan, Osaka, 1955; Abstract, p 45; cf. Kitahara, Y. *Sci. Rep. Tohoku Univ., Ser. 1,* **1956,** *39,* 265.

113. Doering, W. von E.; Knox, L. H. *J. Am. Chem. Soc.* **1954,** *76,* 3203. See also Dewar, M. J. S.; Pettit, R. *J. Chem. Soc.* **1956,** 2021, 2026.

114. Nozoe, T. *Prog. Org. Chem.* **1961,** *5,* 132–165.

115. Nozoe, T. *Kagaku to Kogyo (Tokyo)* **1967,** *20,* 64–71.

116. Ikemi, T.; Nozoe, T.; Sugiyama, H. *Chem. Ind. (London)* **1960,** 932.

117. Nozoe, T.; Takahashi, K. *Bull. Chem. Soc. Jpn.* **1965,** *38,* 665.

118. Nozoe, T.; Takahashi, K. *Bull. Chem. Soc. Jpn.* **1967,** *40,* 1473; Takahashi, K. ibid. 1462.

119. Special issue dedicated to T. Nozoe, which includes the complete list of Nozoe's papers until then: *Heterocycles* **1978,** *11,* 1–58.

120. Koch, H. P. *J. Chem. Soc.* **1951,** *108,* 512.

121. Kuratani, K.; Tsuboi, M.; Shimanouchi, T. *Bull. Chem. Soc. Jpn.* **1952,** *25,* 250.

122. Tsuboi, M. *Bull. Chem. Soc. Jpn.* **1952,** *25,* 369.

123. Schimanouchi, K.; Sasada, Y. *Acta Crystallogr. Sect. B* **1973,** *29,* 81.

124. Furukawa, K.; Sasada, Y.; Shimada, A.; Watanabe, T. *Bull. Chem. Soc. Jpn.* **1964,** *37,* 1871.

125. Sasada, Y.; Oseki, K.; Nitta, I. *Acta Crystallogr.* **1954,** *7,* 113, and following papers.

126. Sasada, Y.; Nitta, I. *Acta Crystallogr.* **1956,** *9,* 205.

127. Nozoe, T.; Seto, S.; Takeda, H.; Morosawa, S.; Matsumoto, K. *Sci. Rep. Tohoku Univ., Ser. 1* **1952,** *36,* 126, and following papers.

128. Nozoe, T.; Sato, M.; Matsuda, T. *Sci. Rep. Tohoku Univ., Ser. 1,* **1953,** *37,* 407.

129. Murata, I. *Bull. Chem. Soc. Jpn.* **1959,** *32,* 841.

130. Nozoe, T.; Sato, M.; Matsui, K. *Proc. Jpn. Acad.* **1953,** *29,* 22, and following papers.

131. Nozoe, T.; Matsui, K. *Bull. Chem. Soc. Jpn.* **1961,** *34,* 616.

132. Nozoe, T.; Matsui, K. *Bull. Chem. Soc. Jpn.* **1961,** *34,* 1382.

133. Doering, W. von E.; Willey, D. W. *Tetrahedron* **1960,** *11,* 183.

134. Nozoe, T.; Mukai, T.; Osaka, K.; Shishido, N. *Bull. Chem. Soc. Jpn.* **1961,** *34,* 1384.

135. Yamakawa, M.; Watanabe, H.; Mukai, T.; Nozoe, T.; Kubo, M. *J. Am. Chem. Soc.* **1960,** *82,* 5665.

136. Nozoe, T. *Chemistry (Taipei)* **1962,** *51,* 156–171.

137. Nozoe, T.; Kitahara, K.; Tezuka, K.; Takahashi, K. Presented at the 15th Annual Meeting of the Chemical Society of Japan, Kyoto, 1962; Abstract, p 235; Kitahara, K. Ph.D. Thesis, Tohoku University, Sendai, 1962; for experimental details, see Nozoe, T. Jpn. Patent 17674, 1964; *Chem. Abstr.* **1965,** *62,* 5234.

138. Nozoe, T. *Kagaku no Ryoiki* **1963,** *17,* 831–849, 1022–1038.

139. Takahashi, K.; Takenaka, S.; Kikuchi, Y.; Takase, K.; Nozoe, T. *Bull. Chem. Soc. Jpn.* **1974,** *47,* 2272.

140. Takahashi, K.; Nozoe, T.; Takase, K.; Kudo, T. *Tetrahedron Lett.* **1984,** *25,* 77.

141. Takahashi, K. *Yuki Gosei Kagaku Kyokaishi* **1986,** *44,* 806–818.

142. Nozoe, T.; Mukai, T.; Murata, I. *J. Am. Chem. Soc.* **1954,** *76,* 3352, and following papers.

143. Nozoe, T.; Seto, S.; Nozoe, S. *Proc. Jpn. Acad.* **1956,** *32,* 472, and following papers.

144. Nozoe, T.; Itô, S.; Kitahara, K.; Ozeki, T. *Bull. Chem. Res. Inst. Non-Aqueous Solutions, Tohoku Univ., Ser. A* **1961,** *10,* 251.

145. Nozoe, T.; Seto, S.; Matsumura, S. *Proc. Jpn. Acad.* **1952,** *28,* 483, and following papers.

146. (a) Nozoe, T.; Ikegami, T.; Itô, S. *Sci. Rep. Tohoku Univ., Ser. 1* **1954,** *38,* 117; (b) Anderson, A. G., Jr.; Nelson, J. A.; Tazuma, J. *J. Am. Chem. Soc.* **1953,** *75,* 4980.

147. Nozoe, T.; Matsumura, S.; Murase, Y.; Seto, S. *Chem. Ind. (London)* **1955,** 1257.

148. Nozoe, T.; Seto, S.; Matsumura, S.; Asano, T. *Proc. Jpn. Acad.* **1956,** *32,* 339, and following papers.

149. Ziegler, K. *Angew. Chem.* **1955,** *67,* 301; Hafner, K. *Angew. Chem.* **1955,** *67,* 301.

150. Hafner, K. *Liebigs Ann. Chem.* **1957,** *606,* 78, and following papers.

151. Nozoe, T.; Seto, S.; Takase, K.; Matsumura, S.; Nakazawa, T. *Nippon Kagaku Zasshi* **1965,** *86,* 346–363.

152. Nozoe, T.; Takase, K.; Nakazawa, T.; Fukuda, S. *Tetrahedron* **1971,** *27,* 3357, and following papers.

153. Nozoe, T.; Takase, K.; Kato, M.; Nogi, T. *Tetrahedron,* **1971,** *27,* 6023.

154. Nozoe, T.; Takase, K.; Nakazawa, T.; Sugita, S.; Saito, M. *Bull. Chem. Soc. Jpn.* **1974,** *47,* 1750.

155. Takase, K.; Nakazawa, T.; Nozoe, T. *Heterocycles* **1981,** *15,* 839.

156. Nozoe, T. Presented at the lst Japan–U.S. Science Seminar in Physical Organic Chemistry, Kyoto, 1965.

157. McDonald, R. N.; Richmond, J. M. *J. Org. Chem.* **1975,** *40,* 1689.

158. Nozoe, T. Presented as the Presidential Lecture at the 32nd National Meeting of the Chemical Society of Japan, Tokyo, 1975.

159. Nozoe, T.; Kawahito, S.; Kimura, A. Presented at the 33rd Annual Meeting of the Chemical Society of Japan, Fukuoka, 1975; Abstract II-618; to be published.

160. Nozoe, T.; Kawase, J. Presented at the 9th Symposium on the Chemistry of Nonbenzenoid Aromatic Compounds, Sendai, 1976; Abstract 2N19; to be published.

161. Nozoe, T.; Yoshida, Y. Presented at the 12th Symposium on the Chemistry of Nonbenzenoid Aromatic Compounds, Matsumoto, 1979; Abstract 1N13.

162. Yang, P.-W.; Yasunami, M.; Takase, K. *Tetrahedron Lett.* **1971,** 4275, and following papers.

163. Nozoe, T.; Yang, P.-W.; Wu, C.-P.; Huang, T.-S.; Lee, T.-H.; Okai, H.; Wakabayashi, H.; Ishikawa, S. *Heterocycles* **1989,** *29,* 1225.

164. (a) Nozoe, T.; Wakabayashi, H.; Ishikawa, S.; Wu, C.-P.; Yang, P.-W. *Heterocycles* **1990,** *31,* 17; (b) in press.

165. Nozoe, T.; Takekuma, S.; Doi, M.; Matsubara, Y.; Yamamoto, H. *Chem. Lett.* **1984,** 627; Matsubara, Y.; Takekuma, S.; Yokoi, K.; Yamamoto, H.; Nozoe, T. *Chem. Lett.* **1984,** 631; *Bull. Chem. Soc. Jpn.* **1987,** *60,* 1415.

166. Takekuma, S.; Matsubara, Y.; Yamamoto, H.; Nozoe, T. *Bull. Chem. Soc. Jpn.* **1987,** *60,* 3721; Matsubara, Y.; Takekuma, S.; Ibata, K.; Yamamoto, H.; Nozoe, T. *Nippon Kagaku Kaishi* **1987,** 1555; Takekuma, S.; Matsubara, Y.; Yamamoto, H.; Nozoe, T. *Nippon Kagaku Kaishi* **1988,** 157; Takekuma, S.; Matsubara, Y.; Matsui, S.; Yamamoto, H.; Nozoe, T. *Nippon Kagaku Kaishi* **1988,** 923; Matsubara, Y.; Morita, M.; Takekuma, S.; Yamamoto, H.; Nozoe, T. *Nippon Kagaku Kaishi* **1990,** 67.

167. Matsubara, Y.; Takekuma, S.; Yamamoto, H.; Nozoe, T. *Chem. Lett.* **1987,** 455.

168. Takekuma, S.; Matsubara, Y.; Yamamoto, H.; Nozoe, T. *Bull. Chem. Soc. Jpn.* **1988,** *61,* 475; Matsubara, Y.; Matsui, S.; Imazu, K.; Takekuma, S.; Yamamoto, H.; Nozoe, T. *Nippon Kagaku Kaishi* **1989,** 1753.

169. Matsubara, Y.; Matsui, S.; Takekuma, S.; Yamamoto, H.; Nozoe, T. *Nippon Kagaku Kaishi* **1988,** 1704; **1989,** 838; Matsubara, Y.; Matsui, S.; Takekuma, S.; Quo, Y. P.; Yamamoto, H.; Nozoe, T. *Bull. Chem. Soc. Jpn.* **1989,** *62,* 2040; Matsubara, Y.; Morita, M.; Matsui, S.; Takekuma, S.; Yamamoto, H.; Ito, S.; Morita, N.; Asao, T.; Nozoe, T. *Bull. Chem. Soc. Jpn.* **1990,** *63,* 1845.

170. Li, M. K. W.; Scheuer, P. J. *Tetrahedron Lett.* **1984,** *25,* 4707, and references cited therein.

171. Nozoe, T.; Ishikawa, S.; Shindo, K. *Chem. Lett.* **1989,** 353. See also Nozoe, T.; Shindo, K.; Wakabayshi, H.; Kurihara, T.; Ishikawa, S. *Collect. Czech. Chem. Commun.* **1991,** *56,* 991.

172. Machiguchi, T.; Hasegawa, T.; Ono, M.; Kitahara, Y.; Funamizu, M.; Nozoe, T. *J. Chem. Soc., Chem. Commun.* **1988,** 838.

173. (a) Nozoe, T.; Seto, S.; Ebine, S.; Itô, S. *J. Am. Chem. Soc.* **1953,** *73,* 1895. (b) Katsura, S.; Sato, K.; Akaishi, K.; Nozoe, T.; Seto, S.; Kitahara, Y. *Proc. Jpn. Acad.* **1951,** *27,* 31, 36, 102.

174. Nozoe, T.; Kitahara, Y.; Masamume, S. *Proc. Jpn. Acad.* **1953,** *29,* 17.

175. Nozoe, T.; Takase, K.; Kitahara, Y.; Doi, K. Presented at the 132nd National Meeting of the American Chemical Society, New York, 1957.

176. Nozoe, T.; Takase, K.; Kawabe, N.; Asao, T.; Yamamoto, H. *Bull. Chem. Soc. Jpn.* **1983,** *56,* 3099, and references cited therein.

177. Yamomoto, H.; Hara, S.; Inokawa, S.; Nozoe, T. *Bull. Chem. Soc. Jpn.* **1983,** *56,* 3106.

178. Nozoe, T.; Takase, K.; Saito, H.; Yamamoto, H.; Imafuku, K. *Chem. Lett.* **1986,** 1577.

179. Nozoe, T.; Ando, T.; Imafuku, K.; Yin, B.-Z.; Honda, M.; Goto, Y.; Hara, Y.; Andoh, T.; Yamamoto, H. *Bull. Chem. Soc. Jpn.* **1988,** *61,* 2531.

180. Nozoe, T.; Takase, K.; Yasunami, M.; Ando, M.; Saito, M.; Imafuku, K.; Yin, B.-Z.; Honda, M.; Goto, Y.; Hanaya, T.; Hara, Y.; Yamamoto, H. *Bull. Chem. Soc. Jpn.* **1989,** *62,* 128.

181. Imajo, S.; Nakanishi, K.; Roberts, M.; Lippard, J.; Nozoe, T. *J. Am. Chem. Soc.* **1983,** *105,* 2071.

182. Zask, A.; Gonnela, N.; Nakanishi, K.; Turmer, S.; Imajo, S.; Nozoe, T. *Inorg. Chem.* **1986,** *25,* 3400.

183. Davis, W. M.; Roberts, M. M.; Zask, A.; Nakanishi, K.; Nozoe, T.; Lippard, S. J. *J. Am. Chem. Soc.* **1985,** *107,* 3864.

184. Nozoe, T.; Kitahara, Y.; Takase, K.; Sasaki, M. *Proc. Jpn. Acad.* **1956,** *32,* 349.

185. Nozoe, T.; Asao, T.; Takahashi, K. *Bull. Chem. Soc. Jpn.* **1961,** *34,* 146.

186. Nozoe, T.; Asao, T.; Takahashi, K. *Bull. Chem. Soc. Jpn.* **1966,** *39,* 1980 and 1988.

187. Fukunaga, T. Presented at the 23rd IUPAC Congress, Boston, 1971; Abstract, p 103, and private communication.

188. Nozoe, T.; Okai, H.; Someya, T. *Bull. Chem. Soc. Jpn.* **1978,** *51,* 2185.

189. Nozoe, T.; Someya, T. *Bull. Chem. Soc. Jpn.* **1978,** *51,* 3316.

190. Nozoe, T. *Pure Appl. Chem.* **1982,** *54,* 975.

191. Nozoe, T.; Someya, T.; Okai, H. *Bull. Chem. Soc. Jpn.* **1979,** *52,* 1156.

192. Someya, T.; Okai, H.; Wakabayashi, H.; Nozoe, T. *Bull. Chem. Soc. Jpn.* **1983,** *56,* 2756.

193. (a) Nozoe, T.; Okai, H.; Wakabayashi, H.; Ishikawa, S. *Chem. Lett.* **1984,** 1145. (b) *Bull. Chem. Soc. Jpn.* **1989,** *62,* 2307.

194. (a) Okamoto, Y.; Honda, S.; Yuki, H.; Nakamura, H.; Iitaka, Y.; Nozoe, T. *Chem. Lett.* **1984,** 1149. (b) Harada, N.; Uda, H.; Nozoe, T.; Okamoto, Y.; Wakabayashi, H.; Ishikawa, S. *J. Am. Chem. Soc.* **1987,** *109,* 1661.

195. Nozoe, T. *Chemistry (Taipei)* **1983,** *41,* A43–68.

196. (a) Nozoe, T.; Okai, H.; Wakabayashi, H.; Ishikawa, S. *Chem. Lett.* **1988,** 1589. (b) Nozoe, T.; Shindo, K.; Ishikawa, S. *Chem. Lett.* **1988,** 1593.

197. (a) Nozoe, T. *Heterocycles* **1990,** *30,* 1263–1306. (b) Kurihara, T.; Ishikawa, S.; Nozoe, T.; Aihara, J. Presented at the 55th Annual Meeting of the Chemical Society of Japan, Fukuoka, 1987; Abstract 3U33; to be published.

198. Nozoe, T.; Wakabayashi, H.; Ishikawa, S. *Heterocycles* **1989,** *29,* 1005.

199. (a) Shindo, K.; Ishikawa, S.; Nozoe, T. *Bull. Chem. Soc. Jpn.* **1985,** *58,* 165. (b) *Bull. Chem. Soc. Jpn.* **1989,** *62,* 1158.

200. (a) Nozoe, T.; Wakabayashi, H.; Ishikawa, S. *Heterocycles* **1989,** *29,* 1459. (b) *Heterocycles* **1989,** *29,* 733. (c)Shindo, K.; Wakabayashi, H.; Ishikawa, S.; Nozoe, T. Presented at the 61st Annual Meeting of the Chemical Society of Japan, Yokohama, March 1991. Abstract 3A 446, to be published.

201. Shono, T.; Nozoe, T.; Maekawa, S.; Kashimura, S. *Tetrahedron Lett.* **1988,** *29,* 555; *Tetrahedron Lett.* **1990,** *31,* 895; Shono, T.; Nozoe, T.; Yamaguchi, Y.; Ishifune, M.; Sakaguchi, M.; Masuda, H.; Kashimura, S. *Tetrahedron Lett.* **1991,** *32,* 1051; Shono, T.; Nozoe, T.; Maekawa, H.; Yamaguchi, Y.; Kametake, S.; Masuda, H.; Okada, T.; Kashimura, S. *Tetrahedron* **1991,** 593.

202. Crombie, L. *Chem. Ind. (London)* **1978,** 663.

203. *Outline of Japanese Geography—Volume of Formosa;* Kaizosha: Tokyo, 1930.

Index of Autographs

Index

Production: Peggy D. Smith
Copy editing: A. Maureen Rouhi
Indexing: Colleen P. Stamm
Acquisition: Robin Giroux

Printed and bound by Maple Press, York, PA